"Deeply felt . . . Rush captures nature with precise words that almost amount to poetry; the book is further enriched with illuminating detail from the lives of those people inhabiting today's coasts. . . . Elegies like this one will play an important role as people continue to confront a transformed, perhaps unnatural world."   —*New York Times*

"The book on climate change and sea levels that was missing. Rush travels from vanishing shorelines in New England to hurting fishing communities to retracting islands and, with empathy and elegance, conveys what it means to lose a world in slow motion. Picture the working-class empathy of Studs Terkel paired with the heartbreak of a poet."
                              —*Chicago Tribune* (Best Ten Books of 2018)

"Sea level rise is not some distant problem in a distant place. As Rush shows, it's affecting real people right now. *Rising* is a compelling piece of reporting, by turns bleak and beautiful."
                    —Elizabeth Kolbert, author of *The Sixth Extinction*

"A sobering, elegant look at rising waters, climate change, and how low lying areas and the vulnerable people who live in those areas are at risk." —Roxane Gay, author of *Hunger*, via Goodreads

"Rush's innovative, brave *Rising* [is] about the changing coastlines of America in a time of climate breakdown, and part of a growing wave of what might be called Anthropocene non-fiction, seeking to find a form for the challenges of our epoch…. [*Rising*] will stay long with me."   —Robert Macfarlane, author of *Landmarks*

"A smart, lyrical testament to change and uncertainty. Rush listens to both the vulnerability and resiliency of communities facing the shifting shorelines of extreme weather. These are the stories we need to hear in order to survive and live more consciously with a sharp-edged determination to face our future with empathy and resolve. *Rising* illustrates how climate change is a relentless truth and real people in real places know it by name, storm by flood by fire."

—Terry Tempest Williams, author of *The Hour of Land*

"Lovely and thoughtful . . . Reading [*Rising*] is like learning ecology at the feet of a poet." —*Minneapolis Star Tribune*

"With tasteful and dynamic didactic language, [Rush] informs the layperson about the imminent threat of climate change while grounding the massive scope of the problem on heartfelt human and interspecies connection." —*Los Angeles Review of Books*

"Moving and urgent . . . Rush's *Rising* is a revelation. . . . The project of *Rising*, like the project of Matthew Desmond's Pulitzer Prize–winning *Evicted*, is to draw attention to ongoing material crisis through the stories of the people who are surviving within it. *Rising* is a clarion call. The idea isn't merely that climate change is here and scary. There's a more important message: There are people out here who need help." —*Pacific Standard*

"Rush traffics only sparingly in doomsday statistics. For Rush, the devastating impact of rising sea levels, especially on vulnerable communities, is more compellingly found in the details. From Louisiana to Staten Island to the Bay Area, Rush's lyrical, deeply reported essays challenge us to accept the uncertainty of our present climate and to consider more just ways of dealing with the immense challenges ahead." —*The Nation*

"Timely and urgent, this report on how climate change is affecting American shorelines provides critical evidence of the devastating changes already faced by some coastal dwellers. Rush masterfully presents firsthand accounts of these changes, acknowledging her own privileged position in comparison to most of her interviewees and the heavy responsibility involved in relaying their experiences to an audience. . . . In the midst of a highly politicized debate on climate change and how to deal with its far-reaching effects, this book deserves to be read by all." —*Publishers Weekly* (starred review)

"A strange new kind of travel guide, *Rising* is a journey through the turbulent forefront of climate change—the coastal communities, rich and poor, human and nonhuman, that are already feeling the first effects of our rising seas. Rush sets out to put a face on a subject that is all too often depicted in abstract graphs and statistics, and gives us a group portrait of the men and women who are fighting, fleeing, and adapting to the terrible disappearance of the land they live on." —Charles C. Mann, author of *1491*

"In this moving and memorable book, the voice of the author mingles with the voices of people in coastal communities all over the country—Maine, Rhode Island, Louisiana, Florida, New York, California—to offer testimony: *The water is rising*. Some have already lost their homes; some will lose them soon; others are studying or watching or grieving. Though they haven't met each other, their commonality forms a circle into which we are inexorably pulled by Rush's powerful words."
—Anne Fadiman, author of *The Spirit Catches You and You Fall Down*

"A poetic meditation on the nature of change, on how people can make peace with a changing world and our agency in it . . . *Rising* [offers] pulsing, gleaming prose and a stubborn search for, if not hope, then peace in the face of disaster."
—*Shelf Awareness*

"Rush rises. She brings stories out of the woodwork, revealing the true effect of sea level rise on the land, on the sea, and on people. . . . Rush makes real a monolithic subject often too large to digest. You can taste the coming salt."
—Craig Childs, author of *The Animal Dialogues*

"*Rising* is not just a book about rising sea levels and the lost habitats and homes—it's also a moving rumination on the rise of women as investigative reporters, the rise of tangible solutions, the rise of human endeavor and flexibility. It is also a rising of unheard voices; one of the eloquent beauties of this book is the inclusion of various stories, Studs Terkel–style, of those affected most by our changing shoreline. A beautiful and tender account of what's happening—and what's in store."
—Laura Pritchett, author of *Stars Go Blue*

"From the edges of our continent, where sea level rise is already well underway, Rush lays bare the often hidden effects of climate change—lost homes, lost habitats, broken family ties, chronic fear and worry—and shows us how those effects ripple toward us all. With elegance, intelligence, and guts, she guides us through one of the most frightening and complex issues of our time."
—Michelle Nijhuis

RISING

RISING

DISPATCHES *from the*

NEW AMERICAN SHORE

ELIZABETH RUSH

MILKWEED EDITIONS

First paperback edition, published 2019 by Milkweed Editions
Printed in the United States of America
Cover design by Mary Austin Speaker
Cover photo © Michael Christopher Brown/Magnum Photos
Author photo by Stephanie Alvarez Ewens
22 23 24 25 26   9 8 7 6

978-1-57131-381-2

Milkweed Editions, an independent nonprofit publisher, gratefully acknowledges sustaining support from the Alan B. Slifka Foundation and its president, Riva Ariella Ritvo-Slifka; the Ballard Spahr Foundation; *Copper Nickel*; the Jerome Foundation; the McKnight Foundation; the National Endowment for the Arts; the National Poetry Series; the Target Foundation; and other generous contributions from foundations, corporations, and individuals. Also, this activity is made possible by the voters of Minnesota through a Minnesota State Arts Board Operating Support grant, thanks to a legislative appropriation from the arts and cultural heritage fund For a full listing of Milkweed Editions supporters, please visit milkweed.org.

Library of Congress Cataloging-in-Publication Data

Names: Rush, Elizabeth A., author.
Title: Rising : dispatches from the new American shore / Elizabeth Rush.
Description: First edition. | Minneapolis, Minnesota : Milkweed Editions, 2018. | Includes bibliographical references.
Identifiers: LCCN 2017059870 (print) | LCCN 2018016302 (ebook) | ISBN 9781571319708 (ebook) | ISBN 9781571313676 (hardcover)
Subjects: LCSH: Coast changes--Climatic factors--United States. | Coasts--Climatic factors--United States. | Sea level--Climatic factors--United States. | Rush, Elizabeth A.--Travel--United States.
Classification: LCC GB459.4 (ebook) | LCC GB459.4 .R86 2018 (print) | DDC 551.45/70973--dc23
LC record available at https://lccn.loc.gov/2017059870

Milkweed Editions is committed to ecological stewardship. We strive to align our book production practices with this principle, and to reduce the impact of our operations in the environment. We are a member of the Green Press Initiative, a nonprofit coalition of publishers, manufacturers, and authors working to protect the world's endangered forests and conserve natural resources. *Rising* was printed by Sheridan Books, Inc.

For my family, for Felipe, and for my scattered tribe.

CONTENTS

——————————————— PART THREE: Rising

Attention is prayer.

SIMONE WEIL

Within a single human existence things are disappearing from the earth, never to be seen again. In Passamaquoddy [Maine] our sacred petroglyphs—those carvings in rock that were put there thousands of years ago—are now being put under water by the rising seas. We've seen this happen for a long time—this diminishing of our natural resources—through climate change and invasive species. The losses have been slow and multigenerational. We have narrowed our spiritual palettes and our physical palettes to take what we have. But the stories, the old stories that still contain a lot of these elements, hold on to the traditional. For example, our ceremonies and language still include the caribou, even though they don't live here anymore. Similarly, we know the petroglyphs still exist, but now they're underwater. The change is in how we acknowledge them.

JOHN BEAR MITCHELL
Penobscot scholar and member of the Penobscot Nation in Machiasport, Maine

RISING

The Password

Jacob's Point, Rhode Island

I HAVE LIVED IN RHODE ISLAND FOR ONE WEEK WHEN I SET out to explore the nearest tidal marsh, the landscape I know will be the first to show signs of sea level rise. I bike across the Washington Bridge, past the East Providence wastewater treatment plant, the Dari Bee, and the repurposed railway station, through Barrington to Jacob's Point. As expected, out along the Narragansett Bay, a line of dead trees holds the horizon. Some have tapering trunks and branches that fork and split. Bark peels from their bodies in thick husks.

The local Audubon ecologist tells me that they are black tupelos. I roll the word in my mouth, *tupelo*, and cannot put it down. *Tupelo* becomes part of the constellation of ideas and physical objects that I use to draw up my navigational charts—I aim toward *tupelo*. Words can shuttle us around in time and space from New England to old England, from Rhode Island back over two thousand years to when the Wampanoag and Narragansett first harvested shellfish in these tide-washed shoals, to a time when language tangibly connected the physical world and the world on the page and in our conversations. Take tupelo, for instance. It is Native American in origin, and comes from the Creek *ito* and *opilwa*, which, when smashed together, mean "swamp tree." Built into the very name of this plant is a love of periodically soaking in water. Word of tupelos once told marsh waders what kind of topography to expect and also where to find relatively high ground.

A month or two before I witnessed my first dead tupelo, and right before I packed up my apartment in Brooklyn and moved north, I found a scrap of language in an essay on Alzheimer's and stuck it to my computer monitor, thinking it might serve some future purpose. It read, "Sometimes a key arrives before the lock." Which I understood as a reminder to pay attention to my surroundings. That hidden in plain sight I might discover the key I do not yet know I need, but that will help me cross an important threshold somewhere down the line. When I see that stand of tupelos I instinctually lodge their name in my mind, storing it for a future I do not yet understand.

Chance has sent me to Providence, but the move feels deeply fortuitous. Here, I think, I will become immersed in the subject matter that has begun to obsess me: the rate at which the ocean is rising. No state (save Maryland, and only by a hair) ranks higher in the ratio of coastline to overall acreage. It is no surprise, then, that 15 percent of Rhode Island is classified as wetlands—and of that 15 percent, roughly an eighth is tidal, both one of the most nimble types of ecosystem in the world and one of the most imperiled. Over the past two hundred years, Rhode Island lost over 50 percent of its tidal marshes to the filling and diking that come with development. Today the remaining fields of black needlerush and cordgrass are beginning to disappear thanks to higher tides and stronger storms.

When I first learned that I would move back to New England in 2015, I also felt a little sick. I grew up seventy-five miles north of here, as the crow flies, in a small seaside community split down the middle between those who came from centuries of money and those who worked in the industries the wealthier residents controlled. My midwestern parents and I were neither. We lived on the nice side of town, but I was the only kid in my neighborhood to go to public school. When I hung out at the private beach I always felt I laughed too hard, that my body moved too wildly. I can still remember one mother loudly telling my own, "Elizabeth plays awfully rough."

"If you have a problem with her behavior you can speak with her directly," my mother responded, gesturing to the water's edge.

Even though I have spent more time in this region than any other place on the planet, coming back didn't feel exactly like coming home. In part because the New England of my childhood is not the New England I encounter now.

In the mornings I ride down the path lining the Narragansett Bay to Jacob's Point just to look at that stand of dead trees. I secure my bike to a wooden fence, then walk across the width of the marsh to shoot black-and-white photographs of their ghostly silhouettes. The trees' bare limbs twine and reach, a testimony to the energy once spent searching for light. I picture the shade they used to cast and the bank swallows awash in that balm, diving like synchronized swimmers, one after another, from the lowest branches.

Or at least that is how I imagine it once was—before the ice sheets started sloughing into the sea, before the shoreline started to change its shape, before the tupelos along the East Bay started to die.

Three years earlier, I'd inadvertently become interested in sea level rise while working on a magazine article about the completion of the longest border fence in the world, which separates India and Bangladesh. As it turned out, the fence was a technicality; people bribed their way through. Water was the real problem. Over the previous fifty years, upstream irrigation projects had diverted over half of the Ganges River's flow. Meanwhile, the Bay of Bengal was seeping into the empty space left behind. Together these two factors led to widespread crop failure.

I will never forget walking the dusty spine of a *char*, a river island formed by sedimentation, behind a boy named Faharul. It took us two hours just to reach his patch of failing mustard greens. A decade earlier this area had been considered one of the most fertile in the region. Now the sere land cracked open, each fissure lined with the white of dried salt. Faharul and I were 150 miles from the coast, and yet what little food he raised often wilted. If the vegetables he depended upon to survive had not carried a trace of salt in their veins he would not necessarily have known that sea levels were rising, and that he himself was vulnerable to this faraway phenomenon. Faharul spoke of the possibility of pulling up his own roots and leaving his family land. His cousin had already fled to India.

I understood then that sea level rise was not a problem for future generations. It was happening already, exacerbated by human interventions in the landscape. And perhaps even more importantly, I sensed that the slow-motion migration in, away from our disintegrating shorelines, had already begun.

My article on the border fence contained none of this. I didn't have the word count, and I was reluctant to play into one of the earliest climate change clichés, that of a drowning Bangladesh. Instead I tucked the knowledge away and returned to the United States. But I was changed, haunted. I had begun to be able to see what those whose lives are in no way dependent upon the coast could not—the early signs of the rise. I found myself reading an unfathomably large planetary phenomenon written into the limp spines of Faharul's mustard plants. Inscribed into the skeletal tupelos at the farthest edge of Jacob's Point.

There is a word coastal landscape architects use to describe a tree that has died due to saline inundation: *rampike*. According to *Random House Dictionary*, the word especially refers to those trees with bleached skeletons or splintered trunks, those undone by natural forces. The word itself is resurrected from an older and slightly more arcane English. A glossary from 1881 spells it *raunpick*, and gives the definition as "bare of bark or flesh, looking as if pecked by ravens." Bare indeed—how exposed and plain, the gesture these trees make alongside our transforming shore.

＊

My first summer in Rhode Island, I return to the marsh often. One morning someone else is there. When he and I cross paths I ask, as nonchalantly as possible, if he knows why these tupelos are all dead. I am trying to find out whether he can see what I can, that the precious balance between salt water and fresh that once defined this tidal wetland has been upset.

"No," the man says, binoculars jangling around his neck. "I'm sorry."

I'll be the first to admit that before I started coming to Jacob's Point I couldn't tell the difference between black tupelo and black locust, between needlerush and cordgrass. I would learn their names only after I realized the ways in which their letters on my lips might point toward (or away from) incredible loss. Then I became fascinated. Because unlike Descartes, I believe that language can lessen the distance between humans and the world of which we are a part; I believe that it can foster interspecies intimacy and, as a result, care. If, as Robin Wall Kimmerer suggests in her essay on the power of identifying all living beings with personal pronouns, "naming is the beginning of justice," then saying *tupelo* takes me one step closer to recognizing these trees as kin and endowing their flesh with the same inalienable rights we humans hold.

Sometime during the last half century, these tupelos' taproots started to suck up more salt water than they had in the past. They were stunned and stunted. Then they stopped growing. The sea kept working its way into the aquifer, storms got stronger and dumped more standing water into marshes, and tupelos all along the East Coast died. Now they no longer bathe the edges of Jacob's Point in shade. The green coins of their leaves are gone, and a recent bird census carried out in Rhode Island's East Bay suggests that the bank swallows are going too.

I tell the stranger all of this. The sentences unspooling fast like the outgoing tide while he shifts from foot to foot, anxious to break away. He has, he tells me, never heard of the tupelo tree. Instead of the luscious rasp of growth on growth and the electric trill of a songbird in flight, out here, at the farthest end of Jacob's Point, we are surrounded by the ticking sound of unprecedented heat. Above us the tupelos' empty, oracular branches groan.

The oldest living black tupelo in the United States sprouted 650 years ago. That means its first buds burst while the plague was killing off approximately one-third of Europe. Now it is the tupelo's turn to succumb in great numbers. And the red knot's. And the whooping crane's. And the salt marsh sparrow's. Of the fourteen hundred endangered or threatened species in the United States, over half are wetland dependent.

Five times in the history of the earth nearly all life has winked out, the planet undergoing a series of changes so massive that the overwhelming majority of living species died. These great extinctions are so exceptional they even have a catchy name: the Big Five. Today seven out of ten scientists believe that we are in the middle of the sixth. But there is one thing that distinguishes those past die-offs from the one we are currently constructing: never before have humans been there to tell the tale. The language we use to narrate our experience in the world can awaken in us the knowledge that transformation is both necessary and ongoing. When we say the word *tupelo* we begin to see that both the trees themselves and the very particular ecology they once depended upon are, at least where they are rooted, gone.

Sometimes a key arrives before the lock. Now I am thinking, sometimes the password arrives before the impasse. These words, when spoken or written down, might grant us entry into a previously unimaginable awareness—that the coast, and all the living beings on it, are changing radically.

❀

One day I decide to visit the Audubon Environmental Education Center at Jacob's Point. It is noon and I am red faced, my shins sliced by bull and catbrier, from spending my morning batting around the dead tupelo. The blue-haired volunteer behind the desk looks at me as though I am mad for having been in the marshes instead of in the air-conditioning, looking at dioramas of the marshes. "Can you tell me about Jacob's Point and those trees at the far end that are dying?" I ask. She suggests I walk through the interpretive exhibit. She even waives the five-dollar fee.

I snake through five rooms where the rhythmic lick of water melting into mudflats sounds from a pair of Sony speakers. The mallards don't move because they have been stuffed with wool.

The box turtles swim tight circles in a tiny tank at the back of a room without windows. I emerge from a papier-mâché cave (a cave in a marsh?) and repeat my question. This time she refers me to Cameron McCormick, the groundskeeper and the person most likely to know what is actually happening at Jacob's Point.

Cameron doesn't have voice mail, so I leave a message with the center's secretary. Two days later he calls me, and we meet at the path down to the marsh the following morning. His eyes are wild and attentive, filled with flecks of cornflower and amber. He wears carpenters' work boots that have come undone and a poorly tie-dyed Audubon T-shirt clearly abandoned by a summer camper. He will spend the rest of the day cutting down invasive head-high grasses called phragmites. Cameron has a degree in ecology and has been managing Jacob's Point for the past five years. It's a process that has become increasingly difficult as the system inputs—temperature, saltwater levels, tidal highs and lows—all shift. He makes a plan, the salt water inundates a new portion of the marsh, and the entire ecosystem changes.

Together we make a beeline for the shore, where Cameron delivers a plastic box full of fishing nets to a group of excited eight-year-olds who are about to catch fiddler crabs. Next we walk toward the stand of tupelos. At first we stick to the high ground. Then, abandoning the idea of keeping our feet dry, we leave the path behind and sink into the soaked land.

Jacob's Point, like all tidal marshes, contains three distinct zones: low marsh, high marsh, and an upland area at its farthest inland edge. Every day the low marsh is covered in salt water twice, and also uncovered twice; the high marsh slips beneath the salt only in storms. Which is to say, along the point's seaward edge, plants and animals have adapted to live with the tides, while upland the opposite is true. Think of a tidal marsh as—like all wetlands—a transitional region where distinctions blur and the entirely wet world morphs into the almost entirely dry one. It is a liminal ribbon. An in-between. A spit of land at the edge of things, where the governing laws change four times a day. Tidal marshes are frontiers, and as Gary Snyder says, "A frontier is a burning edge, a frazzle, a strange market zone between two utterly different worlds." To pass from one to the other is to cross an almost imperceptible but important boundary, the place where freshwater meets the brine of the sea.

As we walk toward the tupelos we are slowly grading downward, crossing the threshold between sweet water and salt. Cameron tells me what he sees and also what he does not see. "These weren't here five years ago," he says, clomping through a bunch of coarse-toothed marsh elders that have taken over a section of the point that has become suddenly rich in saline. "I expect more are on their way, but it's hard to keep up with." The knee-high shrubs have pushed out a stand of phragmites, their arrival making Cameron's job easier in this small acre. But the equilibrium they have brought is not destined to last.

"In the past, when sea levels dropped, the marsh dropped down too, and when they rose the marsh rose with them," Cameron says as we work our way past the tupelos toward the rugosa-studded bank. If you were to take an aerial time-lapse photo of the process he is describing, it would look as if Jacob's Point and the ocean were moving in and out together, the way desire follows the desired.

This swirling, migratory dance is primarily the result of two different physical and ecological processes. The first is called accretion. "As salt water flows in and out of the marsh, vegetation traps some of the sediments suspended in it, and as those sediments settle the marsh gradually gains elevation," Cameron tells me. Accretion results in the building up of low-lying land; it is nature's nimble backhoe. If accretion makes marsh migration possible, then rhizomes power the retreat. Dense, arterial, and interconnected, these specialized root systems run belowground, giving wetlands their shape. In the past, as sea levels rose and the marsh gained sediment, rhizomes would pull away from the increased salinity while simultaneously sending out new shoots, often uphill, in search of the kind of water that suited them best. As these plant communities moved up and in, the fauna that depended on them moved too. While the physical location of the salt marsh might change, its defining features would not.

But now that sea levels are rising faster than they have in the last twenty-eight centuries, the ocean and the tidal marsh are falling out of sequence. In the Ocean State, and along the rest of the Atlantic coast of North America, the rate of the rise is significantly higher than the global average. Here accretion is already being outpaced, which means that land that once was built up slowly is starting to slip beneath the sea's surface. On top of that, if the marsh's upland slope abuts some piece of human infrastructure—a road, or, as is the case at Jacob's Point, an old railway line—as the rhizomes pull away there is nowhere less salty for them to thrust their spindly roots. The marsh is squeezed between the sea and the hard stop we built along its upland edge and, like the tupelo, it begins to drown in place.

"Maybe if the old Bristol line weren't there, Jacob's Point would stand a chance. But then again maybe not. It's so hard to tell with accretion rates being what they are," Cameron says. Then he adds, "It's a terrifying and wonderful time to do the work that I do."

Together we walk over the farthest bank, toward the shade of a bushy beach grape. Out in the Narragansett Bay, a flotilla of sailboats tacks back and forth, working its way south. The boats are from the nearby Barrington Yacht Club, which runs a summer program for locals. From where I stand it looks as though the campers can't be much older than ten or twelve. "Capsize!" their instructor suddenly bellows, and all the little white sails dip beneath the surface of the bay. All, that is, but one. Then I hear another voice, whimpering, "I'm scared of being in over my head." You and me both, kid.

♦

That fall I attend the Rising Seas Summit at the Sheraton Hotel in Boston. I am there on a Metcalf Institute Fellowship, designed to deepen the relationship between environmental writers like me and the scientists on whom our work depends. Over the past twenty years, sea level rise modeling has gotten increasingly sophisticated. Today many modelers rely on a mixture of observational data (such as tidal gauge records), theorems that take into account the earth sciences (for example, factoring in the gravitational pull an ice sheet will have on a nearby body of water), and the geologic record (which provides insight into how quickly sea levels rose in the past). What nearly everyone agrees on is that sea level rise is accelerating at a rate far faster than modern man has ever witnessed. But precisely how high the waters *will* get, especially at any particular location and time in the future, remains somewhat difficult to predict. Among other things, sea level rise is not uniform. As ice sheets melt, their gravitational pull lessens and the ground beneath them rebounds, lowering sea levels nearby while simultaneously intensifying the phenomenon elsewhere. In other words, the places farthest from the largest chunks of melting ice, including the East Coast of the United States, are likely to experience higher rates of relative rise.

The ceiling of the Sheraton is covered in strings of crystal globes like dewdrops and ten-foot-wide linen lanterns likely dyed far away, in the cadmium-ripe fabric factories I've seen polluting the rivers around Dhaka. Beneath those looming lanterns and ridiculous lights I listen to all the different and conflicting predictions being made about the future of sea level rise. I hear about the two and a half feet of rise predicted by 2100—and the ten feet of rise predicted by 2100. The eighteen feet of rise if the planet heats another two degrees—and the thirty feet of rise if the planet heats by the same two degrees. I hear that, counterintuitively, the melting of the West Antarctic Ice Sheet will affect sea levels in Rhode Island more than the melting in Greenland—and I hear about the president of Kiribati buying land in Fiji because his island nation will soon be underwater.

"It is not a question of if but when," says Ben Strauss, sea level rise expert and vice president of the nonprofit news organization Climate Central. Then he shows us a series of photorealistic mock-ups of the world's major coastal cities, starting with Boston. "This image illustrates what four degrees of warming would look like, and corresponds roughly to business as usual," he continues. "Business as usual" assumes that we will emit slightly more CO_2 in the next eighty years than we have since the beginning of the Industrial Revolution.

On the screen above Ben's head, light-blue water washes over just about everything in the city except Beacon Hill and the northernmost corner of Boston Common. MIT is underwater. So are Northeastern University, the Museum of Fine Arts, Fenway Park, Copley Square, and Newbury Street. That was the street where I bought my first Ani DiFranco album. The street my father revitalized in the late eighties by leasing town houses to European bistros and that above-mentioned record store. The street that transformed our formerly working-class family—my grandfather sold seltzer door-to-door—into one with the opportunities privilege provides. The street that paid for my college education.

On the screen, that street is gone.

Then Ben switches to a rendering that shows the maximum two degrees Celsius of warming the Intergovernmental Panel on Climate Change (IPCC) recommends to avert catastrophe. The intersection of Massachusetts Avenue with Huntington Avenue—the very spot where Ben Strauss is standing—is still buried under a layer of blue.

In the ethereal hyperlight of the conference center I see that no matter what we do, many of the landmarks we have long navigated by are going to disappear. *It is not a question of if but when.*

During that fall I begin to suffer from an acute form of anxiety. Nameless storms so large they leave my house lightless and full of water spin into my dreams. My faith in natural processes, in the intricate systems of reciprocity that I was raised to believe keep nature from tilting out of balance, is lost. Gnawing uncertainty takes its place. I wonder if there is a threshold between immersing myself in my subject matter and drowning in it, and whether I have crossed that line. At night unprecedented storm surges rearrange the furniture and my family lineage. The commonly held notion that what has happened will happen again, that there are no new stories, this idea becomes fat with water, fully saturated, then it too slips beneath the sea's dark surface.

✳

Whenever I can, I pull away from my computer screen and ride back out to Jacob's Point. There I wander in a landscape we do not yet have a name for, a marsh inundated by too much of the very thing that shaped it. I have read about the disappearance of tree frogs in Panama, the droughts scraping across Kenya, the heat waves killing thousands in Paris and Andhra Pradesh and

Chicago and Dhaka and São Paulo. I have written about communities affected by sea level rise. But *my* life has seemed so removed, so buffered from those events.

At Jacob's Point I am finally glimpsing the hem of the specter's dressing gown. The tupelos, the dead tupelos that line the edge of this disappearing marshland, are my Delphi, my portal, my proof, the stone I pick up and drop in my pocket to remember. I see them and know that the erosion of species, of land, and, if we are not careful, of the very words we use to name the plants and animals that are disappearing is not a political lever or a fever dream. I see them and remember that those who live on the margins of our society are the most vulnerable, and that the story of species vanishing is repeating itself in nearly every borderland.

In a hundred years none of these trees will be here. No object thick with pitch to make the mind recollect. And if we do not call them by their names we will lose not only the trees themselves but also all trace of their having ever been. Looking at the bare tupelos at the farthest edge of Jacob's Point, I am reminded of something John Bear Mitchell said when my students asked him how the Penobscot people of Maine have responded to centuries of environmental change. "Our ceremonies and language still include the caribou, even though they don't live here anymore. . . . The change is in how we acknowledge them." His response surprised my students. He seemed to be saying: learn the names now, and you will at least be able to preserve what is being threatened in our collective memory, if not in the physical world. His faith in language clearly eclipsed their own.

✻

And then there is the pleasure of it. I like my excursions best when I am alone. Waking early to ride to a slender little marsh that most overlook. The wild blackberries, ripe from summer heat, seemingly fruiting just for me. The black needlerush dried in logarithmic spirals, and patches of salt marsh cordgrass that look like jackstraws and blowdowns in an aging forest. Both bearing the delicate trace of the last outgoing tide.

Beyond the stand of tupelos, the marsh still hums with the low-grade sound of honeybees hunting in loosestrife. The ospreys cast their creosote shadows over cicadas and lamb's-quarters and bayberry. This tiny journey into the marsh feels like a grand field trip. Mud snails wrestle in the ebb tide, a great egret hunches at the far horizon scanning for mummichogs, and the sea balm rushes through the tree of heaven. I walk out only a fifth of a mile farther than most people go, and yet there is so much happening, so many unexpected gifts and self-made surprises.

Dropping down, I arrive at the water's edge. I pull on my bathing suit and dive into the bay, but not before stubbing my toe on a barnacle-covered rock submerged just beneath the surface. I care intensely about being here, about coming back alone and often, and I don't really understand why.

Sometimes the key arrives before the lock.

Sometimes the password arrives before the impasse.

Speak it and enter a world transformed by salt and blue.

Say: *tupelo.*

PART I

Rampikes

Persimmons

Isle de Jean Charles, Louisiana

SOMETIME DURING MY FIRST WEEK ON THE LOUISIANA bayou, I walk to the Isle de Jean Charles. The Island Road, built in the early fifties right after the first oil rig went in, runs eight miles southwest from Pointe-aux-Chenes out between two expanses of water so new that neither has a name. Since the little remaining land is incredibly flat, the sky's extravagant clouds serve as a sort of alternative to topography. Hoodoo-shaped cumulus formations hug the horizon, where a storm is fixing to start. Snowy egrets dig in the few remaining bayou banks, and mullet throw themselves out of the water as the first dime-size droplets of rain fall. Less than halfway to the island, my gut confirms what I already know from my research. This is a world unto itself, coming undone.

Just fifty years ago, the surrounding geography was complex and interconnected—a network of lakes and marshes that were navigable in flat-bottomed boats called pirogues. If you didn't have a boat, you could walk between places by sticking to the higher ground abutting the arterial bayous. This word, *bayou*, sounds French, but it is actually Choctaw in origin. It means "slow-moving stream." Today it is used in a general sense to describe Louisiana's rare riparian coast, even though the bayous themselves are disappearing. The natural ridges and pathways that the Choctaws used to travel are going with them. Nearly every defining feature has been replaced by a single element: salt water.

The loss is pronounced enough that a few years ago the National Oceanic and Atmospheric Administration had to remap the nearby Plaquemines Parish and in so doing removed thirty-one place-names. Yellow Cotton Bay, English Bay, Cyprien Bay, Dry Cypress Bayou, and Bayou Long; none of these individual bodies of water exist anymore. The wetlands that once gave them shape have disintegrated, making the bayous and bays indistinguishable from the surrounding ocean.

"Maybe you could swim," the owner of the Pointe-aux-Chenes marina tells me when I ask if I can get to the Isle de Jean Charles without a car. "But I wouldn't, on account of the gators. Better just to take a right off of Highway 665 and stick to the Island Road." Behind him stands a fifteen-foot-high statue of Jesus. The martyr's body is lank and lean, his arms outstretched toward the watery expanse. Next to the statue a dead cypress tree looms. Its empty branches mirror the man's sacrificial gesture. It too has passed beyond the barest version of itself into death. Its roots soaked in salt.

✽

The film *Beasts of the Southern Wild*, a postapocalyptic tale of a band of homesteaders who survive a fierce storm and eke out a living in the drowned world that follows, was shot on the island and is based loosely on the lives of those who still reside out there, many of whom identify as Native American. I remember watching the film and thinking it remarkable that I was seeing environmental destruction bringing a community closer together instead of breaking it apart. I wanted desperately to know what that might look like in real life. This was long before I moved to Rhode Island, long before I saw my first dead tupelo, but after my initial trip to Bangladesh. It was the summer of 2013 and I was looking for proof of the rise in the United States, so I flew to Louisiana.

When Benh Zeitlin, the director of *Beasts*, told me that the island "felt like the end of the world," I wasn't sure if he was speaking of its remote location or of something less literal. The farther I amble out the single-lane highway to Jean Charles, the more I realize that both explanations make sense. The Isle de Jean Charles is where North America's immense solidity ends, the frayed fingers of fine tidal lace splaying seaward. It is also possible to catch a glimpse of the future out here, of a world where the ocean covers what we used to think of as the coast. That is because over the past sixty years the wetlands that once surrounded the Isle de Jean Charles have all drowned, the rate of accretion trumped by land subsidence, erosion, and sea level rise. When I squint, it is difficult to tell just where the Island Road ends and where the water begins.

A man in a black pickup slams on his brakes and rolls down his window. "There's gonna be some rain. Need a ride?" he asks, leaning back against the cab's cracked leather and pulling at the brim of his baseball cap.

"I'm just walking out to the island," I answer. This doesn't clarify matters for him, so I add, "I'm OK," and shake open my umbrella. He shrugs, rolls up his window, and keeps driving in the other direction, back toward Houma and firmer ground.

I repeat the phrase *I'm OK* in my mind as I walk along the rock-lined road. Three nights before flying to Louisiana, I fled the apartment I shared with the man I was to marry. For months I had sensed that this was not the relationship that would buoy me through the long passage that is adulthood, but I resisted leaving because there was still love, if fraught, between us. Eventually the levee of my optimism broke, and I stuffed my rolling suitcase with clean underwear, empty notepads, and a tent. It is not the first time, nor the last, that I turn my back on something I care about immensely. Though it is the first time it feels like a form of deceit.

Soon the road makes a sharp left along the highest and most stubborn spine of land. Two miles long and a quarter mile wide, this is what remains of the Isle de Jean Charles. Less than half a century ago, the island was ten times larger. Waterfowl marshes surrounded this chenier, or wooded ridge, atop which hundreds of residents built their lives. Now many of the homes that flank the Island Road sit on sixteen-foot-high stilts. Briars billow from the windows of those remaining on the ground, undoing the frames one growing season at a time. The ratio looks like one-to-two: for every lifted house there are two abandoned ones. For every person who has stayed, two are already gone.

Out toward the island's tip, a man sits underneath his raised home enjoying the storm wind, backlit in the bruised light. As I walk past he hollers, "Get off this island!" Behind me a minivan crawls by, and the driver cackles from her window. "You leave!" she heaves in response. I am relieved that neither is speaking to me. The wheels of her Toyota Previa crunch across broken bits of blacktop as she pulls into the driveway next to one of those deflated swimming pools that look like big blue doughnuts. Now, I decide, is as good a time as any to start endearing myself to the people who still live out here. I turn and head toward the house, cantilevered up over the surrounding remnants of marsh. Much to my surprise the seated man says, "You must be Elizabeth."

"Well, then, let me guess. You're Chris Brunet," I reply as I step onto the poured concrete slab that serves as his porch. He is the only islander who returned the calls I made in the weeks leading up to my trip. We had hatched a plan to have lunch on Friday, before I got it in my mind to walk out to the island a day ahead of schedule.

"I wasn't expecting you out here until tomorrow." Chris braces his arm against the seat of an adjacent wheelchair, throws his weight forward, and twists into place. He was born with cerebral palsy but it hasn't slowed him down much. We shake.

"I must be pretty lucky to run into you," I say. "I haven't seen anyone else on the island all afternoon."

"I don't tend to go far," Chris replies. A horsefly circles his head like a ball on a string. "And you just missed Theo, down the way. I saw him drive past in his pickup a bit ago." He must have been the man who offered me a ride.

The woman in the van is Chris's sister Teresa. She shakes my hand and walks over to the refrigerator, where she unloads an armful of soda pop, sweetened tea, and bottled water.

"I'm starting to get a feel for the place. It's awfully pretty," I say.

"Even prettier at sunrise and sunset," Chris adds, pushing up the sleeves of his cotton baseball jersey. "You can't say nothing about this here island until you see both. When it lights up the sky—putting the clouds in different colors—well, I don't know how much you'd pay to see that on a vacation somewhere." He picks up a yellow vinyl chair and rolls over to offer me a seat.

Chris's nephew Howard—whom Chris took in a few years back, along with Howard's sister, Juliette—is fishing in the channel behind the house. In 1951 the first oil rig was installed nearby, and with the rig came "channelization," the digging of access routes through the marsh. The oil companies were supposed to "rock" each channel—to backfill it—when the rigs left, reducing the movement of water through the fragile marshland that surrounds and supports the bayous. "But they didn't do that, they didn't maintain the bayou like they said they would, and now the gulf is at our back door," I was told in town. Every year, thanks to erosion, the channels grow wider, eating into the land that once comprised Jean Charles.

Just then, a dolphin swims up the man-made waterway, past the spot where Howard is casting his line. For a second I find its undulating fin thrilling.

"Forty years ago you would have never seen that animal all the way up here," Chris says. "But the land is opening up all around us. The cuts they made in the marsh speed up the process. What was once sweet water is now salt, so these dolphins, they come in." The entire time that humans have inhabited these bayous, it would have been unimaginable to find a big marine mammal so far "inland." Then again, this island isn't inland anymore.

While the dolphin is not direct evidence of sea level rise, its sudden appearance does point to a dramatic shift over time. It is a shift that Chris, who has lived here his entire life, can perceive and that I, as a visitor, cannot. My initial thrill settles into some other emotion. I think if I look at that dolphin long enough, fix my gaze on its body nosing north, I will once again be able to see—as I did with Faharul's limp mustard plants—what is largely invisible to anyone who moves as often as I do: the hallucinogenic transformation of our coastline, the salt water kneading in—into the aquifer and root systems, into our backyards and basements and wildlife refuges and former freshwater creeks—a change so large it unsettles our very ideas of who we are and how we relate to the land we have long lived atop.

"Is there some single thing that you saw that made it clear to you that the environment was changing, and not just in a normal way?" I ask Chris. "Like that dolphin. When did you see your first one?"

"I don't know, maybe fifteen years ago," he says, scratching his graying goatee. "But you have to know that the thing that makes this saltwater intrusion so damaging now is that it's close to home. Yes, we get dolphins up here, but the worst destruction is taking place in our communities." He pauses, unsure of how to go on. Teresa fills the silence by offering me an Arizona iced tea.

It seems to me that what Chris can't quite put his finger on is that the dolphin is just a symbol. For me the symbol is environmental; I can point to it and say, *This is evidence that the ecosystem is in flux*. But for Chris the dolphin represents the slow disappearance of his neighbors. Over the past forty years nearly 90 percent of the islanders have moved inland. When the people who long called this place home left, they took a little piece of Chris's own idea of home with them.

The dolphin heads back down the channel. It likely encountered some piece of riverine infrastructure, a floating barge, levee, or floodgate. All of which were put in place to protect Houma, the parish seat, from the storms that seem to come at a rate of once or twice a year now.

The dolphin swims past the homes with their roofs blown off. Past the molding mattresses, and the trailers with their piping ripped out. Past the gas pipeline that broke during Gustav and was never repaired, leaving the residents without heat in the winter. Past the empty firehouse. The hundreds of dead cypresses and oaks. And the fishing camps destroyed by Rita. Past Theo's parents' old home, and Lora Ann's old home, and Albert's old home, and all of the other residences that have been abandoned because rebuilding is tiresome and expensive. So tiresome and so expensive that for some, leaving Jean Charles became the best option in a set of only bad options.

I have started to think that those who lived on the island and fled are some of the first climate refugees. By 2050 there will be two hundred million people like them worldwide, two million of whom will be from right here in Louisiana. And then there is Chris, who stays.

"Mind you," Chris says as if he were reading my thoughts, "there is no real difference between those who go and those who stay. After a while people left because of the challenges of living here." His eyes are bright and damp and his skin slick. "When a hurricane hits, you have no bed, no sofa, no lights, no icebox, no gas, no running water, sometimes no roof for a month or more. You sleep on the floor, if you have that to go back to, and you start to rebuild. Or you leave."

"I see," I say, looking up at the floor of his house, which hovers overhead.

Teresa takes this as her cue, hugs Chris, and heads back to her minivan.

"It's not that those who left wanted to go. But each person has a decision that only they can make. And if you are one of those who left, there is still a big part of you that wants to be here," Chris says.

I think for a moment about my apartment in Brooklyn with its view of the S train that I will likely never see again. Then I think of the man I left inside it, whose presence has defined much of my life for the last three years.

Chris watches my features turning inward in the waning light, but is, I think, too polite to pry. Sensing the importance of making myself vulnerable too, I offer the information up, try to turn the interview into an exchange of ideas between equals. When I tell Chris about my flight, about my personal life imploding, his warmth deepens. "Child," he says, "you have to do what is in your heart, even when it's hard. But if he's taking energy from you, then you know what you need to—" Chris looks out toward the open water, his voice trailing off.

In that moment I cease to be the reporter from far away and instead become a mirror in which he can test out and analyze the causes and consequences of leaving someone or something you love. As I watch a series of unknowable thoughts rearrange his owl-like face, I realize that my attachment to my former fiancé, to my apartment, and to the vision of the future I have spent the last couple of years conjuring is much less fierce than Chris's to this island. This shrinking strip of land that, for fifty years, he has rarely left. If it was hard for me to choose to give up a life I had imagined and invested in, what, I wonder, would it take for Chris to let go of the only place he has ever really known?

Chris invites me to visit the next day, and I accept. I walk back down the Island Road, and every hundred yards or so, I pass a huge cypress or oak stripped bare, its leafless branches reaching like electricity in search of a point of contact. The cause of the trees' untimely demise isn't in the air, but deep in the ground where the roots wander, where the salt water has started to work its way in. Just south of the Island Road, half the trees have fallen into the widening channel. Those that are still standing are just barely so. Everything, it seems, leans toward the salt water that wasn't always here.

*

On my way back down Highway 665, I stop to buy some groceries at the Pointe-aux-Chenes Supermarket, a low-slung building with a long white veranda and a limited selection of shrink-wrapped vegetables in Styrofoam packaging. Inside, a woman speaks with the cashier about the squashed snake she almost stepped on in the parking lot. I check on the snake—a garter—and notice a bumper sticker on her rusting Camry's trunk. The state of Louisiana is bright yellow, and inside it are the words "Shaped like a BOOT because we kick ASS."

The irony is that Louisiana isn't shaped like a boot anymore. Back at my rental house in Montegut, I pull up an aerial picture of the state on Google Earth. Today the wetlands that once made up the boot's sole are all tattered and frayed. They look more like mesh than rubber. And in fifty years they are likely to be gone entirely. According to the United States Geological Survey, Louisiana lost just under 1,900 square miles of land between 1932 and 2000, an area roughly equal in size to Delaware. And it is likely to lose another 1,750 square miles by 2064, an area larger than my soon-to-be-adopted home state of Rhode Island.

That's because the southern edge of Louisiana is eroding at a rate among the fastest on the planet, and sea level rise and the oil industry aren't the only things to blame. The Mississippi River is directly responsible for building up the coast of the Bayou State. For much of the past ten thousand years, it deposited silt from the far reaches of the continent here, where it emptied into the sea. The world's fourth-longest river drained a vast watershed stretching from Wyoming to Pennsylvania, from the border of Canada to the Gulf of Mexico. In wet years a section of the river that might typically be one mile across can swell to as many as fifty (as has happened all along the river's lower reaches in present-day Arkansas, Mississippi, and Louisiana), picking up additional soil and sediment and carrying it south.

Pre-Columbian Native American societies understood that a healthy river goes through cycles of flood and drought, and they shaped their civilizations around the Mississippi's ebb and flow.

Their villages were sited not on the banks but nearby, and most weren't permanent settlements but camps that could be relocated if the waters rose. In 1543, however, the Spanish conquistador Hernando de Soto was stopped in his march westward across present-day Tennessee by a swollen Mississippi. His chronicler, Garcilaso de la Vega, mentions the encounter in his book *The Florida of the Inca*; it was the first time (to the best of my knowledge) that the Mississippi's regular high waters and sediment-delivering surges were described as a deterrent to human progress. The second recorded instance of the river's "wrath" came in 1734, when it flooded a fledgling New Orleans. Then in 1927 the river inundated an area the size of Massachusetts, Connecticut, Rhode Island, and Vermont combined for several months, destroying new towns that had sprung up all along its banks. It wasn't until the Mississippi got in the way of the colonial project that its predictably fickle flow was deemed a problem.

In an effort to "manage" the mighty river, the Army Corps of Engineers put in one dam, then two, then three, then nineteen. Today there are twenty-nine dams and locks on the upper Mississippi, and the lower Mississippi is lined with levees and floodwalls. Instead of preserving the low-lying land at the Mississippi's mouth, these river controls have contributed to its destruction by impounding land-replenishing sediment behind man-made barriers upstream. Thanks in part to these interventions, the Isle de Jean Charles, and the wetlands surrounding it, started to disappear, not just temporarily beneath floodwaters, but for good.

✱

The next day I drive back out to the island for my midday meeting with Chris. Here on the far reaches of the bayou, my visit is an event. Another one of Chris's nephews, Dalton, comes over to watch *Mission: Impossible* and when the movie ends he joins our conversation. The afternoon is hot and still. The three of us sit together and eat slices of store-bought cake from the market in Pointe-aux-Chenes. The talk moves easily through a range of subjects: the kids' schooling, the bus schedule, the weather. It is a great comfort to be engulfed in the workings of a family not my own.

I felt immediately at home despite the fact that Chris's house is physically falling apart. The plaster and particleboard have been stripped from all the walls, and the bones of the structure shine through. In order to save it from mold after Hurricane Lili, in 2002, Chris gutted the entire thing.

"That Lili, she got all the way into the house here," he says with a sweep of his arm. "I had to take out all the walls. I've been repairing them little by little, but the going's slow." He rolls from the living room to the kitchen and offers me a soda. "When more people lived on the island I would have been able to call on some of them and get help with this here," he continues. The bedsheet tacked between his bedroom and the kitchen flaps in the wind. "Now I get help occasionally but mostly do the work myself, one board at a time." His living room, which is separated from the children's bedroom by a faded piece of red fabric, has been under construction for more than a decade.

"Do you remember," Dalton says, "I forgot what hurricane it was, when they were dropping all them sandbags from the helicopters? You know that levee busted for the fourth time during the storm and they still haven't finished fixing it."

"It wasn't Rita, and it wasn't Gustav or Katrina or Ike," Chris says, rattling off names with an ease that borders on the familial. He looks out the window to where the nameless bay laps at the disappearing land and laughs. "If they really wanted to save the island they would have included it in the Morganza to the Gulf protection plan." Chris is speaking of a $13 billion infrastructure project to construct ninety-eight miles of levees that would wrap most of the Terrebonne and Lafourche parishes in ten-foot-high earthen berms. The project, which is part of an even larger so-called Master Plan designed to "rescue" part of the state's crumbling coast, will require $50 billion to complete. That's more than the costs of the Manhattan Project, the recovery from Sandy, and the Hoover Dam combined.

"No one is surprised that we weren't among those who were saved," Dalton says firmly. "We are Indians, after all."

On the last morning of my trip I speak with Albert Naquin, the reigning chief of the Biloxi-Chitimacha-Choctaw tribe to which Dalton and Chris belong. Albert, like so many others, doesn't live on the island anymore. He moved to Pointe-aux-Chenes—a stone's throw from the grocery—after he had to mop an inch of mud off brand-new appliances and dining room furniture in his first year of marriage. "I was fresh out of the army with a baby on the way. The first time I flooded, it was the end for me," Albert tells me, tugging at a black baseball cap with the word *NATIVE* embroidered on it in big block letters.

Albert, who is in his sixties and built like an old Buick, has spent the last twenty years trying to organize the remaining islanders to relocate as a group and to get the Army Corps of Engineers to pay for it. While Chris says he isn't against the idea, he has yet to wholeheartedly embrace it. And others are completely opposed. Back in 2002, when the initial Morganza to the Gulf feasibility report was submitted and Jean Charles left out, that was as close as Albert ever came to uniting the islanders.

"I think the Army Corps was feeling guilty about not including us in their big plan, so they offered to help us relocate," Albert tells me. "But we needed to show that nearly everyone living on the island would be interested in leaving. On the day we met with the government folks there were a bunch of people who don't even live on Jean Charles asking all of these questions that derailed the conversation. After that the interest in relocation dropped, and without consensus no one was going to give us money to move."

Before coming to Jean Charles I researched the history of Louisiana's wetlands. Not surprisingly, our knowledge of early residents is somewhat limited; most artifacts have been found in less ecologically volatile areas upstream, such as the Cahokia Mounds of Illinois. The Chitimacha are said to have lived in what is present-day central Louisiana for over six thousand years. In the face of the violence that accompanied the arrival of Europeans, they migrated south along the lower Mississippi in the eighteenth and nineteenth centuries, arriving at the far reaches of the delta at roughly the same time as the Biloxi and the Choctaw, who were retreating from their ancestral homes in the wake of Florida's bloody Seminole Wars.

The convergence of so many disparate Native groups—along with the Acadians, who were expelled from Nova Scotia and other soon-to-be-Canadian provinces by the British in 1755—on the boggy fringes of the continent was no coincidence. Living in this marshland—considered uninhabitable by most mainland Europeans—was a kind of shared survival tactic, and Acadians and Native Americans thrived together here. But today the high rate of intermarriage between these groups means that the federal government does not recognize the residents as Natives. And since the island was never formally a reservation, there is no federal mandate to relocate the islanders now that their home is disappearing.

"At first we were losing one or two families with every storm," Albert says. "But now, with the wetlands opening up, the storms are getting worse, and over the years the flow of people off the island has increased. If it continues like this, eventually there won't be anyone left out there. And who we are, our unique Native community, will become fractured, will disappear along with the land."

＊

A light wind moves through Chris's house, making the exposed beams whistle. This place was built by Chris's grandfather, who insisted on using Douglas fir trees for their strength and resistance to rot. It has been standing in exactly this spot for the better part of the last century, though Chris has lifted it twice: first after Hurricane Lili, and then even higher after Katrina.

"For a while my parents were completely self-sufficient," he says, "but by the time we were adults they went to the grocery store." Chris grew up eating blackberries, oranges, pears, and cantaloupes all grown in the garden alongside his home. Back then gardening was easy, because there wasn't any salt in the groundwater.

He rolls over to a big wooden chest and lifts out a warped photo album. I watch as he flips past pictures of his house, water lapping at the window frames, back when the structure rested on the ground. Past black-and-white photos of his father playing guitar. Past the image of himself, much younger, and the brother who died a few years back, orphaning Howard and Juliette. Past the image of his sister Teresa, in a pair of John Lennon sunglasses and a striped jumper, smiling in front of a live oak tree with a double trunk and Spanish moss cascading down. The same one still stands behind the house, but is a husk of its former self today, a rampike with all its branches removed. Past the photo of Father Roch's place next door and the forest of cypresses that once separated *here* from *there*.

Eventually Chris arrives at the photo he wants to show me. In it his father is tilling the ground in a dirty white button-up shirt, flanked by okra plants. "That was all the way back in 1959," Chris says, "the year he married my mother." His father is working the land his parents had given him as a wedding gift. Chris runs his finger over the image and hands it to me. "It looked so different back then."

I have been in the Terrebonne Parish for over a week, and everywhere I go people keep telling me how it used to be. They even have photographic evidence. It is almost as if the islanders have lived on a different island. A near-perfect copy of the Jean Charles of today, but ruled by a slightly different set of laws. Everything here is just as it was there, with a few notable exceptions.

The cypresses are all in the same places, but their leaves have vanished. Some of the land where gardens once sat remains, but salt rests in the soil; the plants won't grow, and the land lies fallow. And what was once a wetland rich in fowl is now open water. In the photo Chris shows me, his father stands surrounded by pastures. You can even make out a black cow in the upper right corner. In the sixty years since, the meadows where the cattle used to graze have all slipped beneath the surface of the sea.

"When I was a boy," Dalton says, "my papa used to go out into the marshes just south of the house. He would be gone all day and would return with a sack full of dead ducks. He gave 'em to people. That's how many ducks he had. My pa was a good hunter, but back then there was also enough to hunt, enough to go around."

Today, if you were to open up Chris's refrigerator, you wouldn't find ducks, fish, beef, or homegrown vegetables. Instead you would probably discover two gallons of industrial milk, three two-liter bottles of no-name soda pop, and a box of Frosted Flakes.

"Right out there, that's where the marshes were," says Chris, pointing south through his paneless window. I look out and see only water. The wind whips up a couple of whitecaps and the sun glitters hard atop each one. "It used to be that you could walk all the way to Montegut without getting your feet wet. Now you can see clear across to the water tower, but you have to take a boat to get there."

Since the ducks that his father used to hunt no longer nest nearby, Dalton drives to Houma to purchase Purdue's saline-soaked poultry. Both he and Chris still eat local shrimp, but they supplement that with government-subsidized grains and vegetables grown by agricultural giants.

"Sometimes we have these unplanned reunions at Walmart," says Chris. "I mean, you can run into a lot of the people who used to live on the island and even those of us that remain. We are all there buying food, catching up. It's nice to see the people I miss."

Chris's statement is so matter-of-fact, so tinged with nostalgia, that I nearly miss its implications. The actions he is describing are not harmless or merely circumstantial; they are a feedback loop, if a relatively slight one. The disappearance of coastal land is causing human beings who were once self-sufficient, whose impact on the planet was slight, to use fossil fuels to procure the food they once were able to grow at home. Every time the islanders drive to Houma they are, in some small way, accelerating the disappearance of this ecosystem. I want to ask if they know the consequences of their new way of life—but I can't think of a way to formulate this question without sounding rude. Instead I ask for another slice of cake.

By the time I return to my rental house, a dark, sinister feeling has taken root. At first I try to distract myself by watching a bad Sandra Bullock movie on television. Then by boiling the shrimp I was gifted back in Pointe-aux-Chenes. When I fail, I step out onto the front porch and watch islands of water hyacinth floating down the channel. Since coming to Louisiana I have temporarily taken up smoking again. I don't know what I hate to admit more, that I smoke three Lucky Strikes out there in the storm light, or that after each one, I cry. For all I left behind and for the even more the islanders have lost. But mostly out of fear. Because I know that the future will look nothing like the past.

✻

Months later I read about a bird the size of a clenched fist. Some people call it the red knot. Others call it the moon bird. That's because it can fly 320,000 miles, or the distance from the Earth to the moon and halfway back again, in a single lifetime. Its migration is one of the longest in the world, stretching between the Arctic and either Patagonia, in Argentina, or Mauritania, in West Africa.

Researchers recently found that the bodies of young moon birds are shrinking because the ice on their arctic breeding grounds melts earlier each year. When the ice melts earlier, the plants bloom earlier, and the insects that eat the plants emerge earlier too, long before the fledgling moon birds are able to feed. Without the nourishment of insect larvae, the juveniles' bodies do not grow to full size. When they fly south, away from the Arctic and the warmth that is made visible in their shrunken feathered wings, they cross the equator and encounter an inescapable truth. Smaller bodies come with shorter beaks.

Because their beaks are shorter, the moon birds are incapable of digging nutrient-rich mollusks from their wetland winter feeding grounds. The hunger of these abnormally small moon birds forces them to gnaw on seagrass rhizomes, which sit closer to the surface. These interconnected root systems are what hold the marine meadows together. They give them shape. And so with each rhizome-packed nibble the moon birds take, the seagrass beds slump a little more, slowly breaking apart beneath the rising tide. Maybe the moon birds will go with them.

I fall asleep with this image floating in my mind: bite by bite, the short-billed red knots unknowingly unknotting the web of their survival.

✦

Chris urges me to visit with Edison Dardar, another of the holdouts, before I leave Louisiana. Edison's home is the first on the left along the Island Road in the community of Jean Charles. Across the way is a small, beached orange submarine from the 1950s. In front of it a handmade sign reads, "ISLAND iS NOT FOR SALE. IF YOU Don't like THE ISLAND STAY OFF. Don't GiVE uP FigHT For YOUR RighTS. It's WORTH SaViNG. Edison Jr."

Chris tells me that Edison doesn't like to speak to reporters. Still, I stop by a couple times during my month on the bayou. No one is ever home, or so it seems. One day, I leave a long handwritten note, introducing myself, mentioning Chris's endorsement, and asking if Edison might allow me to call on him. I hear nothing. On my second-to-last morning on the island, as I am driving back toward Pointe-aux-Chenes and another interview, I decide to pull over, park, and try one last time.

I consider walking up the stairs to the house, but if my time on Jean Charles has taught me anything it's this: most people pass the afternoon beneath their homes or along one of the bayous, where the breeze is strong. I walk across the two-by-fours that connect the road to the concrete slab turned toolshed beneath Edison's moss-green cottage. There are six cast nets hanging from a beam. Five plastic buckets. A washbasin. A scale. Various wrenches and other rusting tools tacked to the stilts that support the house high overhead. Cans of bug spray and Rust-Oleum line the jerry-rigged shelves. There is a laminated poster of Jesus and a pregnant black-and-white cat lounging atop an overturned crate. Miraculously Edison is also there, standing among it all, cradling a single yellow cucumber.

"What's that?" I ask by way of greeting, even though I already know.

Edison looks at me, sighs, and says, "It's yellow, so at least it's good for seed. I get one or two a day. But it ain't nothing like it used to be, with cucumber and green bean vines everywhere and a big garden besides." His voice is low and slow and full of the local drawl I will miss once I am gone. I extend my hand and introduce myself.

"I know who you are," he replies. His blue eyes search mine while the four plastic pinwheels tacked at the front of his workbench whir.

We walk together through the tall grass that surrounds his home and out to a homemade altar. The altar has four levels; each is wider than the last and is fashioned from found wood. On the platforms are all different kinds of objects—antique soda bottles, oyster shells, fishing buoys, rusting crab cages, and even a couple of strands of faded Mardi Gras beads. Presiding over the pyramid of wonders are two duck-shaped hunting decoys, ceremoniously screwed into place. The whole thing reminds me of Simon Rodia's Watts Towers—those dreamy and reverent cantilevered cones that invest meaning in objects that others normally regard as trash. "I was an oysterman, and we were always pulling up strange things from the bayou. If I saw something nice, I would bring it home and add it here," Edison says, his gray hair going haywire in the wind. "But that was forty years ago."

Back when there were more people on the island, men would

gather by the altar, drink a couple of beers, and talk over the daily catch. They might even go down to Antoine Naquin's Dancehall, which was also the church and the five-and-dime, to listen to the local zydeco band. "I get mad when people leave," Edison tells me, lifting objects from the altar. Each, I imagine, triggers a memory. "You know, the more people on the island, the bigger the island."

I let his words sit in the dense air. I know they seem illogical; people can't make the island larger, and in fact its diminishing size is what leads most to leave. But there is truth to what Edison says. After a month of listening to the islanders' stories I have come to think of the community as a kind of organism. The more people there are, the more robustly this organism can organize and reconstitute itself. With more people on the island, post-storm recovery is fast; with more people on the island, gas lines are repaired. With more people on the island, you don't have to drive so far to get what you need.

Edison's father lived all of his ninety-one years on Jean Charles, Edison tells me, and his father's father spent his whole life here too. "We've been down here a long time," he says. "It used to be that you could catch four hundred pounds of shrimp a night in the little inlet right there. I still bring in enough to eat now, but not much more than that." Later he will pull a five-gallon bucket out of his refrigerator, with about fifty shrimp squirming in the bottom. Two big whites and the rest brown, probably no more than two pounds total, or one two-hundredth of what he caught in a day when the fishing was good.

Before we leave the altar he hands me a flare of oyster shells, growing out of and on each other. "When one oyster dies," Edison says, "the next one builds on his shell, and the next one builds on him. Me? I plan on dying right here, on the island." I try to hand the shells back but he refuses. "You take that with you," he says. "A souvenir."

I run my thumb along the shiny inside of a shell, where the

oyster's belly once fit snug. It is a gift the land gave to Edison and that today he is determined to pass on.

Edison is opposed to Albert's relocation strategy. He fears that if the islanders all agree to leave, the land will be sold off to the highest bidder. This anxiety may seem irrational, but it is, at least partially, informed by half a millennium of Western wrongdoing to Native communities all over the Americas. It is also a sentiment that I will encounter regularly in vulnerable coastal communities throughout the United States. Those who have the least are often the most reluctant to give up their small share, especially if others will turn a profit from their sacrifice.

Together we work our way through the roseau cane back toward the house, stopping at what remains of Edison's garden. Instead of planting directly in the ground, he uses a couple of bathtubs that he salvaged from homes abandoned nearby. The plastic barrier keeps salt water out of the roots. There are three cantaloupes and five yellow cucumbers. Out front two persimmon trees Edison planted to replace those lost during Hurricane Andrew swim in the wind.

The long arms of the trees are laden with round fruit ready to fall. The air is heavy with the smell of vegetal ripeness and coming rain. "Each tree you have is good protection during a storm, plus the fruit is pretty tasty too," he says, plucking one from the branch and handing it to me. I think of Li-Young Lee's poem "Persimmons," and recite the first few lines aloud. Edison chuckles.

How to eat:
put the knife away, lay down newspaper.
Peel the skin tenderly, not to tear the meat.

We walk together back to his workbench, and I cup the persimmon in my palms. Lift it up and down to better sense its density. The shiny globe is full of sun and the little freshwater that still snakes its way along the island's stubborn spine. I lift it to my face, breathe in deeply, and smell the land giving of itself to make a musky sweetness. Before digging my nails into the skin I pause, wondering if I shouldn't eat it—there are so few, and the one that I hold in my hands plays no small part in Edison's ability to stay. But not eating it would be rude. So I take a bite, and the fruit's thick pulp runs down my chin, luxurious and strange. It is a taste I have never encountered before. And in that moment I think I know why he and the others do not leave.

On Gratitude

Laura Sewall: Small Point, Maine

I WAS WORKING IN MY HOME OFFICE THE DAY HURRICANE
Bill hit. It was a hot August day. Perfectly calm, perfectly clear. But
I could hear big waves. I looked up from the desk, out the window
and across the marsh, and saw these huge waves crashing on the
other side of the dunes. The water was coming into the marsh really fast because there was suddenly so much of it to move within
the twelve-hour tidal cycle. I ran over to my sister's and she and
her husband were up on the roof photographing these big whirl-
pools swirling in the marsh. It was somehow so magical I jumped
in. But I got scared immediately—and I never get scared swimming. I remember thinking, *This is not any pretty water.*

So I came back here, got my kayak, and paddled out into the
marsh. It was completely, utterly covered. It looked like a big solid
mirror. I remember floating past patches where just a few inches
of grass stuck out. Because the rest of the stalks were submerged,

the tips of the blades were absolutely covered in bugs. As I floated by they tried to jump into the kayak. I realized then that there is so much life in these marshes that is not prepared for higher waters. I mean, I never even thought about the insects. Where are they going to go?

After that I had a real spike of something like fear. I thought if I were to be honest with you I would admit that I don't know what is going to happen to the marsh in front of my house. I don't know whether some big surging wave is going to spill over that little peninsula and come pounding through my windows. I don't know if the marsh will be able to keep up with the rise. I actually think it is years away, but I am not so sure that someone buying my house could say it is a generation or two away. And that is how the houses down here are thought of, in generational terms. So what is scary, in an immediate sense, is that I may not have the retirement funding I thought I had by virtue of selling this house. I don't know what is going to happen with that. I have no uncertainty about the climate science, but I do have a lot of uncertainty about what to do.

I have watched a brand-new pool form on the marsh; I see the land being eroded. Right on that edge over there, eleven feet have been lost since 2004. Some people say I am in denial. But that is a really ineffective and inaccurate way of referring to a particular psychological process. Living here is not denial. It is a choice. I am sixty years old right now. I could watch some really amazing change in the time I have left and I could stay until it gets washed away. I would be, I don't know, say eighty-five by then. It just might be perfect. I don't have kids so I don't need to pass anything down. It is a very self-centered perspective, I know that. But it is not denial.

These decisions are complex because there are a lot of factors to take into account. For one, I have to take into account my incredible love for sitting right here. I feel so privileged to be observing these changes so immediately. It is frightening but it is also incredibly interesting, awesome really. There is something magical and enlivening about seeing how dynamic life is on the planet. You think of animals running around but you don't think of plants moving. See that big patch of brown grass over there? It is migrating uphill because it is not super salt resistant and where it used to be is a relatively low part of the marsh that now gets flooded more often than before. So I am seeing a different kind of grass, *Spartina alterniflora*, come in behind it. I am literally watching the ocean encroaching right here in front of my house and it amazes me.

But there are also nights in the winter when the wind will be blowing so hard I fear that my metal roof is going to rip off and be shredded into pieces that pierce through the windows. This fear drives my spiritual work. Where I go with it, on a personal level, is toward making peace with uncertainty, toward being more fully in the present, and toward living a life where gratitude is near the surface.

I came across this old reminder; do you know Brother David Steindl-Rast? His work has a theme that I love. Essentially, he says that it is not that you can have gratitude for everything all the time but that there is always the possibility of gratitude; there is always something that you can tap into to feel your gratitude, no matter what. Thinking in this way takes care of so much of my anxiety. It is very easy for me to feel gratitude for the place I live, at least when I have time, when I am not consumed by work. There have been too many days in this last year where I was grumpy.

It had nothing to do with where I was but with the fact that I didn't have time to appreciate this place. I was locked into the computer or the tasks. And so many of them are uninteresting. Lately my feeling is that I need time to just be here before I can decide whether to stay or not. My guess is that I will tap into so much gratitude for my life alongside this marsh that I may just become an old lady who drowns right here.

The Marsh at the End of the World

Phippsburg, Maine

A GNARLED OLD PINE MARKS THE ENTRANCE TO THE Sprague River Marsh. It is high summer, a short season of riotous green in Maine. But the tree hasn't taken any cues from the tilting of the planet, the long hours of sunlight, or the sudden warm spike. Its branches extend empty and bare. This pine must be about a hundred years old, but as with so many others I saw lining the banks of tidal marshes up and down the coast, too much salt water had too regularly soaked into the ground around the tree's root system, killing it. On the surface, this single tree might seem inconsequential. But its death is a sign of a much larger transformation—the disintegration of tidal marshes all along the coast, from the Sacramento—San Joaquin River Delta to the Gulf of Mexico.

In the eighties hardwoods and pines often thrived along our marshy shore. Now they do not. It is still hard for me to believe that a departure this big began in my lifetime. I've encountered so many of these rampikes that I have come to think of them as a

series of memorials, a supersize Christo and Jeanne-Claude installation that spans the entire country, from the Louisiana bayou all
the way to this remote corner of the Gulf of Maine. Together they
commemorate the tipping point: the moment the salt water began
to move in. And now that sea levels are rising more quickly than
they have in the last three thousand years, an even bigger change
is happening. The ground itself has begun to rot.

I walk through a patch of poison ivy and over a weathered
outcrop of granite into the marsh. The moment I step onto the
upper portion of the Sprague I know that it is in trouble. There I
am met by the musky, almost strawberry scent of decomposition.
Most marshes smell a little bit, but here the scent is overwhelming. A healthy marsh is firm underfoot. Here the earth quakes like
Jell-O. With every step bubbles burble from the accrued depths,
releasing the captive sulfur that lies beneath.

For the researchers I will visit at the Sprague, the smell of the
rotten marsh is halfway normal. For me it conjures up images of
a neglected compost bin.

In my mind, rot is something vegetables do. The fruit arranged
for a still life will rot, which is why some artists prefer to paint
plastic apples and pears. Limbs rot when gangrenous. I did not
think, until coming to the Sprague, that it was possible for the
ground itself to rot. Or that when it does it might just help heat up
this precious pebble even faster.

❋

"Welcome to our rotten marsh," says Beverly Johnson, a professor of geology at Bates, the small liberal arts college about thirty
miles inland where I too teach.

Beverly speaks a kind of hybrid language—half scientific fact,
half casual like a block-party conversation. Her wardrobe is a

similar mix of business and pleasure. She wears knee-high wading boots, long black shorts, and a maroon T-shirt with a hiker and mountain peak airbrushed across the front. She carries in her periwinkle Osprey pack a change of socks, three water bottles, and a yellow hardcover all-weather geological field notebook, the words *Gulf of Maine* scribbled down the spine in black Sharpie.

Dana Cohen Kaplan and Cailene Gunn, two of Beverly's students, who have been studying the relationship between marsh degradation and the release of greenhouse gases for their senior thesis projects, accompany her in the marsh. Bates students and faculty have been conducting research in and around the Sprague since 1977. Forty years ago their concerns were notably different; one of the earliest theses written about the greater Bates-Morse Mountain Conservation Area investigates porcupines and their food preferences. Today, most who make the journey to the coast study how and why the area is changing as a result of human activity. In our era of unprecedented geologic transformation, the very act of scientific observation has taken on an added sense of urgency. In the coming years, portions of, if not all, places like Jacob's Point and the Sprague are likely to be underwater. We will want to know why, but we need the data first. The chance won't come again.

In recent years scientists have discovered that coastal wetlands—salt marshes, but also mangroves and saw grass meadows—store a quarter of the carbon found in the earth's soil, despite covering only 5 percent of the planet's land area. That means that an acre of healthy coastal wetlands will clean far more air than an acre of the Amazon. "They sequester about fifteen times more carbon than upland forests," Beverly tells me. "But how effective are these ecosystems when they have been dammed, diked, culverted, or drained? That's what we'd like to know."

Dana unloads a large Plexiglas box and an eighty-thousand-dollar machine that looks like a waterproof stereo receiver from the back of the college van. "It's a cavity ring-down mass spectrometer," says Joanna Carey, a biogeochemist who, like the machine, is on loan from the Marine Biological Laboratory at Woods Hole, Massachusetts. "We use it to measure carbon dioxide, methane, and water vapor levels being 'respired' by the marsh so we can get a better idea of how higher sea levels will alter the net balance of greenhouse gases in these already-altered coastal ecosystems." As the marsh is further destabilized, it is possible that the organic matter that was stored in and around the root systems will decompose, releasing back into the atmosphere the very gases—carbon dioxide and methane among them—the marshes once sequestered.

Dana places the contraption into a wheelbarrow. "Cailene and I nicknamed it the Science Box," he says. It used to be that we thought the earth's climate and its underlying geology changed slowly and steadily over time, like the tortoise who beat the hare. But now we know the opposite to be mostly true. The earth's geophysical makeup doesn't tend to incrementally evolve; it jerks back and forth between different equilibriums. Ice age, then greenhouse. Glaciers covering the island of Manhattan in a thousand-foot-thick sheet of ice, then a city of eight million people in that same spot. The transition between the two is often quick and relatively dramatic. Contraptions like the Science Box help us keep track of just how fundamentally things are changing, illuminating the ways in which human activity is pushing the planet beyond "greenhouse Earth" into some even warmer, preternatural state.

The Science Box takes various vapor emission readings at a rate of one per second. From these readings Dana will generate one "flux," or an image of the overall rise or fall in the methane and carbon dioxide coming off one square meter of marsh. Then he will compare the fluxes gathered in healthy areas against those in places that have already begun to rot from within, creating a picture of the potential impact sea level rise will have on a tidal marsh's ability to sequester greenhouse gases.

The amount of data the Science Box generates in four minutes would take a human 3,600 minutes to collect by hand. Which is exactly what Cailene spent her summer doing in Long Marsh, a "fingerling" tidal wetland about ten miles northwest of here as the crow flies. There is no road to the marsh's terminus; to reach the transition zones where the readings are most telling, Cailene must drop down the side of a culvert near the marsh's mouth. Then she hikes through the waist-high grasses, hopscotching across rivulets and drainage ditches until she reaches the end. It takes her thirty-three minutes to travel from stem to stern. She can't safely cart the Science Box all the way back there, which is why she collects her readings the old-fashioned way—with a twenty-five-milliliter syringe and an Exetainer vial. Tapping away at the calculator app on her cell phone, she says that it took her months to produce one-tenth of the data the team will collect today.

Cailene and Dana will devote much of the upcoming academic year to better understanding what separates a healthy tidal marsh from one that is not, and the rate at which each releases greenhouse gases into the atmosphere. Or, as Beverly describes it, "They are filling in the equation that describes *today's* carbon cycle."

As I drove down State Route 209 and out on the fog-struck peninsula that morning, the local NPR radio personality likened the weather to pea soup. The midday heat was bound to break records, he warned. Now, listening to Cailene, I understand that it is going to be not only the hottest day of the summer but also one of the most important, at least for these young researchers. As we prepare to walk, Dana adjusts his straw cowboy hat and tugs at his sun-bleached Cisco Brewers T-shirt, pulling it over his belt.

Then he looks out across the sea of saltwater cordgrass and black needlerush, places his hands on the wheelbarrow handles, and enters the humming midmorning light. Not only will today's work net the raw material of his yearlong thesis project, it will hopefully help illuminate how drowning tidal marsh ecosystems could inadvertently contribute to the ongoing inundation of the coast.

*

For much of human history we have had very little sense of the dynamic nature of life on the planet. Three hundred years ago we didn't know that the earth has been regularly covered in massive sheets of ice that pulse in and out from the poles like a scab forming and retreating. We didn't know that the continents were in constant motion or that animals could go extinct. We didn't know that light traveled faster than sound or that bacteria caused disease, and we didn't know that the universe began not with God's word but with a big bang.

Right up through the middle of the eighteenth century, Westerners thought the earth began roughly four thousand years before Christ. But unearthing evidence of species that modern humans knew absolutely nothing about—such as a massive mastodon molar found in present-day Kentucky—hinted that there had once existed many other worlds, which had flourished and vanished over a previously unimaginable length of time. One of the earliest books to acknowledge the idea that the earth's history might be much longer than our own was Charles Lyell's *Principles of Geology*, written just over a century and a half ago. It popularized the work of William Smith and James Hutton, who spent decades comparing the appearance and disappearance of different fossilized animals in the red sandstone cliffs in Devonshire, England, in the late 1800s. As John McPhee writes in *Annals of* the *Former World*, "Some creatures . . . had appeared suddenly, had evolved quickly, had become both abundant and geographically widespread, and then had died out, or died down, abruptly. Geologists canonized them as 'index fossils' and studied them in groups" in order to get a better sense of the age of our planet. The earth scientists at Devonshire painstakingly compared these "index fossils" against each other and in doing so started to divide geologic time into different epochs. Their studies suggested that, contrary to popular belief, the earth had likely been gyrating just outside the asteroid belt for the better part of four hundred million years.

Of course this estimate of the earth's age was not accurate either. It wasn't until radiometric dating was pioneered by Arthur Holmes at the turn of the last century that we improved on this rough calculation—by a huge margin—and discovered that our planet actually came into existence roughly 4.5 billion years ago. Though our tools have progressed, most nongeologists, me included, are still likely to wildly misidentify different events in geologic time, often by orders of magnitude.

Four thousand, four hundred million, or 4.5 billion years—it is all the same to us. We tend to think in human lifetimes, and even there our scope is limited. We are individually preoccupied by the lives of those we know and expect to know: our grandparents, parents, children, and, if we are lucky, grandchildren. Which is why it is so fantastically difficult for us to recognize that in our frenzied attempt to keep nearly eight billion people fed, watered, clothed, sheltered, and distracted, we are fundamentally altering the geophysical composition of the planet at a pace previously caused only by cataclysmic events, like the massive asteroid that smashed into eastern Mexico, wiping out the dinosaurs, sixty-five million years ago.

Lately, Earth-minded scientific researchers and activists alike have taken to condensing the history of the planet into a single calendar year to explain just how temporally insignificant human civilization is and how profoundly we have changed the planet in the time it takes, relatively speaking, for a rufous hummingbird to beat its wings. In this version of history, the planets are formed at the very beginning of January. Sometime during the first week of the year, a giant object collides with Earth, and out pops the moon. It isn't until late July that the first cells form. In August coral creeps across the ocean floor. Late in October multicellular organisms appear. Plants make their way onto land close to Thanksgiving. Around the first of December come the amphibians and insects. Dinosaurs arrive on December 12, and by December 26 they are gone. On the evening of December 31 the first hominoids emerge in East Africa. At ten minutes to midnight Neanderthals spread to Europe. We invent agriculture one minute before the clock strikes twelve. Shortly thereafter we start to write things down. All it takes is five short seconds for the Roman Empire to rise and fall. We enter the industrial era two seconds before midnight, the petroleum age a half a second before the year comes to a close. And in that fraction of a second we cause the end of an entire epoch.

The Holocene closes and the Anthropocene (or the Capitalocene, as environmental historian Jason W. Moore suggests calling it) begins, launching a geologic period defined by the complete and utter dominance of certain human beings and our endless accumulation of resources. In that fraction of a second, we open the earth's veins, exhume as much energy as possible, and pump various byproducts into the air, causing the atmosphere to warm twenty times faster than normal. We cause the polar ice caps to melt, the oceans to heat, and the coastline to change its shape. We alter the very makeup of the biosphere, the twelve-mile-deep sliver of the earth that is home to all known life that has ever existed in the entire universe. "Abundant" and "geographically widespread" are two ways of describing the extent of humans' impact on the planet. Lately I have been wondering whether the descriptor "index fossil" might also soon apply.

＊

Global sea levels have risen about nine inches since we started keeping track in 1880. If they were to keep rising at this rate, by century's end they would be roughly five inches above where they are today. But most scientists expect to see anywhere between an additional twenty-four to eighty-four inches of sea level rise by 2100, and every year the estimates creep higher still. Between the turn of the last century and 1990, sea levels rose, on average, 1 to 1.2 millimeters per year. Then the rate of the rise itself started to increase, rapidly. In the intervening quarter century, the per-year increase has risen to 4 millimeters, and, like so many other climate change signals, it shows little sign of slowing down.

As the rate of the rise continues to accelerate, tidal marshes are becoming inundated and, as here, they are starting to rot. If you drill into a healthy marsh you quickly encounter a network of rhizomes and black, iron-rich sediment. This sediment, which cements most marshes together, is so dense it doesn't contain any oxygen. And this anoxic environment is, in part, what makes marshes such good carbon sinks—whatever organic matter is stored there decomposes extremely slowly because it is never touched by air. But when salt water sits on a marsh and cannot drain, as is happening in tidal wetlands the world over, the marsh grass rhizomes either retreat or rot.

Some places, like the southern edge of Louisiana and the Isle de Jean Charles, have already passed through this transitional process. Others, like vast swaths of the Everglades, are just beginning to show signs of collapse. As these marshes become flush with salt water, they are contributing to atmospheric warming—but just how much, and at what rate, remains unclear. That's partly because each location is unique, with different kinds of flora respiring at different rates, and also more generally because throughout Western history tidal wetlands were thought to be the homes of swamp serpents and marsh monsters, the boggy, slimy sources of malaria, disease, and death. As such, they have long gone overlooked, which is why the research taking place out here in the Gulf of Maine is so important.

The US Fish and Wildlife Service didn't understand the connection between marsh rot and climate when it decided to "plug" a ditch in the Sprague likely dug by the Civilian Conservation Corps in the early 1930s. The Sprague River Marsh is not unique in this way. By the end of the decade following the Depression, over 90 percent of New England's saltwater marshes were grid-ditched, mostly in attempts to reduce mosquito populations in coastal communities. All along the Eastern Seaboard, workers took shovels to swampy land, hoping to drain the sections prone to retaining water.

The Civilian Conservation Corps didn't care that ditching would transform the hydrology of the entire ecosystem. The standing water in which mosquito larvae hatched was greatly reduced—and with it went hundreds of other species. Dragonflies and water beetles. Mummichogs and silversides. The seaside sparrow. The great egrets and white ibis. So, over a decade ago, the US Fish and Wildlife Service started plugging the ditches. They thought intervening in an already altered hydrological system might be able to return the marsh to a state of equilibrium. They thought they might be able to bring back the water beetles and wading birds. But, it turned out, layering one kind of human intervention on top of another only dragged the Sprague further from its starting point.

Not much more than a four-foot-by-eight-foot piece of ply-wood, a ditch plug is a simple-enough idea: it is meant to stop tidal flow through man-made channels, reintroducing an element of standing water into the marsh. But ditch plugs are too effective at restricting flow. Fresh water from the upland side filters into the marsh and does not continue toward the sea. And whenever an exceptionally high tide or storm surge arrives, breaching the barrier, salt water gets stuck in place there too. As a result everything above the plug is permanently inundated with saline-rich water, and as the water starts to evaporate, the saline concentrations shoot even higher. The rhizomes in the marsh grasses, unused to these conditions, begin to decompose; the ground around them collapses; and the greenhouse gases long stored in the sediment are released into the air. At least that is what these scientists suspect is happening.

"The Fish and Wildlife Service really screwed this up," says Beverly, straddling the channel behind the plug, bloated with brackish water. "Though they know this now." The edges of the plywood in front of her are egg-yolk yellow and dusty green, the center buckled.

Later, when I type "what rots" into Google, the search engine tries to finish my question, suggesting *What rots teeth? What rots first when you die? What rots quickly?* I discover that acid rots teeth. Cell membranes in the liver are the first thing in the human body to rot. When improperly stored, potatoes rot quickly, and I don't need Google to tell me that they smell bad when they do.

Google does not suggest making my sentence *What rots marshes?* It is not the first time the search engine—thanks in part to its millions of users, whose habits dictate the autocomplete option—has been, in my humble opinion, misleading. Because if marshes are among the largest carbon sinks in the world, and if rot transforms them into huge carbon sources, then we surely *do* want to know what rots marshes and, perhaps more importantly, if there is anything we can do to better prepare them for the future that is already here.

When I look out across the white slime that coats the once-loamy ground above the ditch plug, I know that what is happening in the Sprague is, in a very basic sense, what will happen to many of the world's marshes as the height of our oceans continues to climb. My fever dreams of tidal wetlands—and all the species endemic to them—drowning, of our coastlines contracting, and of mass migrations inland return with prehensile force. They drag me deeper into the marsh, out into the rotting cordgrass where the ground quakes like chocolate pudding. There, at the decomposing center of the Sprague, I stand dumbstruck by our planet's transformations.

I am starting to be able to see not just the dead trees sprinkled along the shore like so much confetti, or the fistful of decaying grass I hold in my hand. I am beginning to make out the rough outline of our future coastline. Everywhere that once was a tidal marsh will likely be open water. The words of Ben Strauss echo again in my mind: "It is not a question of if but when."

❋

"We know that healthy marshes have historically kept pace with moderate changes in sea levels, but how they respond to those kinds of changes when ditched, plugged, and tidally restricted is another thing," says Cailene. The two tiny silver geckos tacked to her ears reflect the sun. "And that's important because, for example, of the hundred and thirty-one marshes here in Casco Bay, one hundred and twenty-eight have been altered."

"There are twelve ditch plugs littered throughout the Sprague," Beverly chimes in. "And hundreds throughout marshes up and down the coast."

A recent study released by the National Academy of Sciences predicts that as coastal wetlands continue to be transformed by atmospheric warming, they will release more methane into the air. But what makes a wetland vulnerable may be more complicated than its height and altitude. As Kimbra Cutlip wrote in a recent issue of *Smithsonian* magazine, "How much carbon wetlands take up, how much they release, how quickly soil accumulates . . . are all factors that are intertwined with one another and dependent upon a variety of influences. Like the tugging of one line in a tangled web of ropes, as one loop loosens, another tightens, changing the shape of the whole bundle." When humans interfere with marsh hydrology—by ditching, plugging, draining, diking, culverting, and developing alongside and in these unique landscapes—they are yanking, even severing, the ropes that tie the marsh together.

In the short term, widening the culverts that restrict tidal flow, removing man-made infrastructure—things like ditch plugs and roadways—and reconnecting marshes to the rivers that have long provided the silt that fuels accretion would likely increase these important ecosystems' ability to keep pace with sea level rise. However, as the rate of the rise itself accelerates, what tidal marshes will need more than anything else is space, room to migrate up and in. And, though few want to admit it, providing space will likely mean relocating some of the human communities we have built along the seashore.

Just below the buckled piece of plywood, Dana drops a Plexiglas chamber over a preselected square of healthy marsh vegetation. Joanna, who has spent much of the past year using the ring-down mass spectrometer to calculate net fluxes all around New England, lays the Science Box on two milk crates. She plugs the machine into a set of tubes that connect to the chamber. Then she presses a button and the Science Box begins to whir, almost immediately producing data. Everyone crowds in to look at the stream of numbers scrolling up the screen.

"Right now we aren't seeing any methane emissions, which is what we want," says Beverly. A molecule of methane, one of the most potent greenhouse gases on the planet, can, over the span of a decade, heat the atmosphere eighty-six times faster than a molecule of carbon dioxide. "And the carbon dioxide is dropping too, because the plants are photosynthesizing," she adds. In essence, they have verified what they already know—healthy marshes are good at sequestering and storing greenhouse gases. The data gathered here, below the ditch plug, will serve as a control to measure the rest against. It takes only a few minutes for a heap of healthy cordgrass to become a set of numbers, a kind of bottom line.

After sampling three different areas where the ground is firm and the grass luxuriant, we move the field station back two hundred feet or so, above the ditch plug. The land starts sucking at our boots again, squelching and giving way beneath us as we plod in.

"An alternative name for my thesis might be 'Measuring Marsh Farts,'" Dana jokes as he tries to keep his balance near a particularly pestilent pool covered in brown scum.

As the group prepares the fourth test site, a lanky research technician who hasn't said much all morning points at the hollow of my throat and asks, "What's with that necklace?"

For a second I am confused. I reach up and grasp a silver hexagon hanging on a silver chain, a Christmas present I hadn't taken off since receiving it. "It was a gift," I say. "Why?"

"It looks like a shorthand representation of the atomic structure of benzene." Then he adds in a wooden voice, "It's classified as a carcinogen in California." The research technician breaks out into a wry and knowing smile, claps a hand on my shoulder, and laughs. I have come to adore science-geek small talk almost as much as I enjoy learning about the inner workings of these often-overlooked landscapes. Those who are devoted to tidal marshes are members of the same scattered and idiosyncratic tribe. They are more at home thigh deep in sulfurous mud than they are at the local shopping mall, and increasingly—as they bear witness, if not to the end of *the* world, then certainly to the end of *one* world—their humor has taken a turn for the macabre. "You have to laugh to keep from crying," a geologist in the Everglades once told me.

The cavity ring-down mass spectrometer beeps, a warning that there is humidity in the lines. Benzene Man turns away from me and faces the malfunctioning machine. The crew disconnects and reconnects the hoses. The beeping continues.

"Science," Beverly says over her shoulder. "Winging it every day."

Once the water is cleared from the lines and we have all eaten a snack, the Plexiglas chamber is lowered over another square of marsh grasses. This time nearly half are rotten. For a moment the world goes silent, everyone leaning in toward the bleached-out computer screen. The first reading is 1.55 parts per million of methane, then 1.6 parts per million, then 1.7 parts per million. All the scientists let out a little yelp.

"It's kind of twisted," Beverly tells me, chuckling. "But when we see that methane increase, it's good, in a way, because it means that our hypothesis is at least partially correct."

Just as they supposed, the rotting patch of marsh grass above the ditch plug is contributing more methane and carbon dioxide to the atmosphere than the sample plot of the same size below. Beverly and her students suspect that the water infiltrating the marsh and now impounded by the ditch plug stimulates methanogens to spring into action, breaking down the organic matter the Sprague has long stored. A kind of fermentation follows that causes the marsh to decompose from within while also releasing methane and carbon into the atmosphere at an unprecedented clip. Tug at a couple of ropes and the shape of the whole bundle changes.

"I'm not opposed to the idea of 'monkey wrenching' the ditch plug," says Laura Sewall, an eco-psychologist and the caretaker of the Bates-Morse Mountain Conservation Area, who has joined us for the second half of the morning. Laura is advocating the kind of small-scale act of eco-defense Edward Abbey once encouraged to reestablish healthy hydrological patterns in the American West. While localized interventions of this sort won't do much to stem the threat sea level rise poses to our most vulnerable coastal landscapes, they can help to temporarily preserve a world worth rescuing. Removing the ditch plug surely is a step in the right direction: coaxing saltwater marshes back toward their original hydrology in the hopes that they will be able to, at least in the short term, rise with sea levels as they have in the immediate historic past.

Whether that immediate past is an appropriate analog for the future is an important question to ask. Sea levels are currently rising much faster than was previously predicted. James Hansen, a former NASA scientist who now teaches at Columbia University, recently published a controversial paper that suggests that the rate of the rise will continue to accelerate exponentially in the coming years. So much so that he predicts that by century's end the world's oceans will likely be many meters higher. In which case monkey wrenching the ditch plug isn't likely to save the Sprague. Removing the human infrastructure, and in particular the road that runs along its upland edge, would provide space for migration, and might be the only chance the marsh has to make it into the next century.

Dana stands next to the Science Box and does some rough

calculations on his phone. "It looks like the area above the ditch plug is releasing *significantly* more methane than the area below."

"Methane," Beverly reminds me, "is, generally speaking, thirty times more effective at trapping heat than carbon dioxide, making it the most potent, if short lived, of the world's greenhouse gases."

In that moment my desperation, of the monkey-wrenching sort, gives way to monumental uncertainty. If what is happening right now on the Sprague is also unfolding in impounded tidal marshes the world over, then the likelihood that we will witness widespread marsh collapse goes up. But no one knows whether it will go up by a factor of one or one hundred, because humans have never recorded these kinds of events before.

What we do know is this: each molecule of methane released into the air warms the oceans and the atmosphere, speeding up the rate at which glaciers and ice sheets are melting, which in turn accelerates the rate at which sea levels are rising, which diminishes the chances that a marsh will be able to adapt, raising the likelihood that it will rot and drown instead—which brings us back to the methane readings on that dimly lit screen on the edge of the Sprague: 1.55 parts per million, then 1.6, then 1.7. Another feedback loop closed and amplifying.

✱

After lunch, Laura and I split from the group for an afternoon kayak. We launch from her house, which sits just across the Sprague River on a small mound of land overlooking the marsh. Laura's ancestors were some of the first Europeans to settle permanently along the Gulf of Maine, but she grew up on the other side of the continental United States.

"When my parents got married, they drove west until they hit water," she tells me as we dig our paddles in deep and pass the breakers where the Sprague pours out into the gulf. "Too much family back here."

I pause and watch a line of terns riding the air currents that rise from the waist-high waves as they curl and break. In the sheltered dunes between the beach and the marsh, a handful of piping plovers are beginning to fledge. Where the cordgrass gives way to woods, pitch pines twist along the edge of a slice of gray granite that looks like a whale's back.

The scene reminds me of the opening lines of one of my favorite children's books, Robert McCloskey's *Time of Wonder*. He writes, "Out on the islands that poke their rocky shores above the waters of Penobscot Bay, you can watch the time of the world go by, from minute to minute, hour to hour, from day to day, season to season." As a child I used to camp with my family a hundred miles north of here, on the quiet side of Mount Desert Island. Returning to this rocky coast makes me feel a little as if my life is on repeat, as if what has happened is happening again. Though when I think about the preliminary findings procured on the marsh that morning, I realize that my familiarity and comfort are illusory; the Maine of today is not the Maine of my youth.

Together Laura and I cruise along the offshore spine of Sewall Beach, the largest undeveloped spit of sand in the state. Like the Sewall Woods in nearby Bath, this place is named for her family. You can't throw a stone around here without hitting something tied to the Sewall legacy. It is a history that Laura finally started to embrace about fifteen years ago. After decades away, working on environmental projects around the world, she returned to Maine, and has begun to act as a kind of liaison between this marsh and the surrounding community, carrying news of the environmental changes taking place in the Sprague to the folks those changes will most immediately affect.

"The people who live out here, from Phippsburg all the way to Small Point, they are starting to pay attention," she says. "In part because the road out the peninsula is already flooding during storms. They even formed an advisory board on the 'New Environment' and asked me to be a member. I'm not sure what the committee will be able to achieve, but at least they're asking the right questions about mitigation, marsh migration, the impact on local fisheries, insurance, infrastructure, all that stuff." Laura is easily one of the most well-informed and deeply committed citizens I have encountered since I started writing about sea level rise. She filters every bit of information she receives, every lived experience, through the lens of climate change awareness, and in so doing gives the seemingly cataclysmic a different sheen.

Together we travel between two distinct but continuous realms—the land-bound marsh and the open ocean. Out here the surface of the water is pure glass, spotted occasionally by the passing of a cloud. Every time I pull my paddle from the sea a tiny wave travels outward and dissolves. Something happens as I nose my little boat closer and closer to the blue-on-blue horizon, where water and sky become indistinguishable. I begin to feel as though I am paddling straight into the heart of a Rothko painting, or a landscape where all traces of memory have been wiped away. The sun strikes the bay, filling my vision like a bell, and the morning's worry momentarily disappears.

It takes us about half an hour to reach the Heron Islands, a set of four granite outcroppings approximately a mile from the shore. Laura has never been out this far before in her kayak. As we approach the islands she tells me that the Herons aren't known, despite their name, for wading birds. Twenty feet to my right, the snout of a horse-head seal slowly rises out of the water. He stares up at me and I stare right back, watching the little wakes that radiate out from where his breath hits the sea.

This day is anything but ordinary, I think. Dulse-colored plumes of rockweed rumble beneath our bows as we slide between the largest of the islands, through a slender channel no wider than a school bus. I look down into yet another little universe at the edge of things: the seaweed below waves brilliant maroon, and a couple of rock crabs scuttle sideways. For a long time, Laura and I say nothing at all. Wordlessly, we head back toward the shore. About halfway there she dips her hand into the water and out come the words, "I've never felt it so warm before."

And with that the spell is broken. My hand follows hers, breaking apart the clouds that slide across the surface of the sea. I think of my childhood summering along the Maine coast. The gulf was usually so cold I couldn't bear to stay submerged for more than a second. Now, as I look down at my fingers comfortably wriggling below, I realize that this too has changed.

These days all it takes is a little unusual warmth to make me feel nauseated. I call this new form of climate anxiety endsickness. Like motion sickness or sea sickness, endsickness is its own kind of vertigo—a physical response to living in a world that is moving in unusual ways, toward what I imagine as a kind of event horizon. A burble of bile rises from my stomach and a string of observations I have been hearing in these parts adulterates the joy of our afternoon adventure. Because the Gulf of Maine is warmer than ever before, the bottom-dwelling cod, pollack, and winter flounder are pulling away from shore. Because the Gulf of Maine is warmer than ever before, the shrimp fishery has been closed for years. Because the Gulf of Maine is warmer than ever before, phytoplankton are disappearing, green crab populations are exploding, and sea squirts are smothering the seafloor. Because the Gulf of Maine is warmer than ever before, the lobster are moving into deeper, cooler waters, keeping the lobstermen and women away from home for longer. Because the Gulf of Maine is warmer than ever before, everyone and everything that lives here is changing radically.

＊

When we arrive back at Sewall Beach, Laura and I throw our exhausted bodies on the hot sand and stare up at the sky.

"We have to become more comfortable with uncertainty," she says, as if reading my mind.

"Those who lived during the plague were probably a little uncertain about their future prospects," I say with a snort. "Maybe we can try to channel them."

Ten feet away, a seagull picks a clam from the surf, flies over the shingle, and lets the shell fall. It drops to the ground, picks the shell back up, and rises then releases it again two, three, four more times.

"For most of human history, mankind hasn't been half as sure of civil order or reliable food sources as we are today," Laura says. "And maybe that sureness isn't such a good thing. Maybe it dulls the senses, makes us less aware of what's happening right in front of us, right now."

Finally the shell the seagull has been struggling with breaks open. A slimy clam belly glistens on the wet sand. The gull calls to a friend and they feast together. For a moment I revel in the beauty of this basic ritual, happening right in front of us, right now. Then I think about how the ocean is, like the marsh, one giant carbon sink. When it absorbs carbon dioxide it becomes more acidic, which makes it difficult for bivalves like clams to build their shells.

"What about those guys?" I ask. I gesture to the seagull duo digging into their lunch.

Laura drags her fingertips through the sand and doesn't answer my question. Instead she squints into the sun, stands, and says, "Let's bodysurf a little before we head in."

And that is exactly what we do. It is this moment that I will remember in the middle of winter, when I wonder whether I made good use of my time, whether I lived fully in the few short months of riotous green here in the northeasternmost corner of the country. We play that afternoon, seal-like in the unusually warm surf. Our bodies held aloft in the curl of a spitting wave, while on the other side of Sewall Beach, salt water sits in the Sprague River Marsh, rotting the land from within.

※

That night as I lie in bed, I remember a Hindu fable about the origins of the universe. It says that every four billion years a flood completely dissolves the earth. Vishnu returns after the deluge in the form of a tortoise. On his back he places Mount Mandara, which serves as a churning rod around which he wraps a snake. Gods and demons grab hold of opposite ends. They tug against each other. The rod turns. The ocean roils, releasing amrita, the nectar of life. And the great earthly dance begins again.

I think then of a perversion of the story popularized during the British colonization of India. It picks up where the original left off and is often recalled as a conversation between an Englishman and an Indian sage.

Question: What does the great tortoise whose back supports the world rest upon?
Answer: Another turtle.
Question: And what supports that turtle?
Answer: Ah, sahib, after that it's turtles all the way down.

I think the exchange is designed to poke fun at the Hindu religion and also at any argument built upon an infinite regression. But I have always been inclined to find some truth in this tale of turtles upon turtles, supporting our earth. When I hear the line "It's turtles all the way down," I don't balk. Humans are nothing more than atoms come together to make life. The things we eat, the air we breathe; it is all made of the same manna. I think of the seagulls and the clams, and wonder what happens to the seagulls when the clams can't make their shells. What happens to Mount Mandara and the sea of milk if the tortoise's back dissolves in an acidic ocean? Perhaps when it dissolves, the world floods and the cycle starts again. Perhaps that is what is happening right in front of us, right now.

Pulse

South Florida

IN 1890, JUST OVER SIX THOUSAND PEOPLE LIVED IN THE damp lowlands of south Florida. Since then the wetlands that covered half the state have been largely drained, strip malls have replaced Seminole camps, and the population has increased a thousandfold. Over roughly the same amount of time the number of black college degree holders in the United States also increased a thousandfold, as did the difference between the average salaries of CEOs and the workers they employ, the speed at which we fly, and the combined carbon emissions of the Middle East.

About sixty of the region's more than six million residents have gathered in the Cox Science Building at the University of Miami on a sunny Saturday morning in 2016 to hear Harold Wanless, or Hal, chair of the geology department, speak about sea level rise. "Only seven percent of the heat being trapped by greenhouse gases is stored in the atmosphere," Hal begins. "Do you know where the other ninety-three percent lives?"

A teenager, wrists lined in aquamarine beaded bracelets, rubs sleep from her eyes. Returns her head to its resting position in her palm. The man seated behind me roots around in his briefcase for a breakfast bar. No one raises a hand.

"In the ocean," Hal continues. "That heat is expanding the ocean, which is contributing to sea level rise, and it is also, more importantly, creating the setting for something we really don't want to have happen: rapid melt of ice."

A woman wearing a sequined teal top opens her Five Star notebook and starts writing things down. The guy behind her shovels spoonfuls of passion fruit–flavored Chobani yogurt into his tiny mouth. Hal's three sons are perched in the next row back. One has a ponytail, one is in a suit, and the third crosses and uncrosses his gray street sneakers. The one with the ponytail brought a water bottle; the other two sip Starbucks. And behind the rows and rows of sparsely occupied seats, at the very back of the amphitheater, an older woman with a gold brocade bear on her top paces back and forth.

A real estate developer interrupts Hal to ask, "Is someone recording this?"

"Yes." The cameraman coughs. "Besides," Hal adds, "I say the same damn thing at least five times a week." Hal, who is in his early seventies and has been studying sea level rise for over forty years, pulls at his Burt Reynolds moustache, readjusts his taupe corduroy suit, and continues. On the screen above his head clips from a documentary on climate change show glacial tongues of ice the size of Manhattan tumbling into the sea. "The big story in Greenland and Antarctica is that the warming ocean is working its way in, deep under the ice sheets, causing the ice to collapse faster than anyone predicted, which in turn will cause sea levels to rise faster than anyone predicted."

According to Marco Rubio, the junior senator from Florida,

rising sea levels are uncertain, their connection to human activity tenuous. And yet the Intergovernmental Panel on Climate Change expects roughly two feet of rise by century's end. The United Nations predicts three feet. And the National Oceanic and Atmospheric Administration estimates an upper limit of six and a half feet.

Take the six million people who live in south Florida today and divide them into two groups: those who live less than six and a half feet above the current high tide line, and everybody else. The numbers slice nearly evenly. Heads or tails: call it in the air. If you live here, all you can do is hope that when you put down roots your choice was somehow prophetic.

But Hal says it doesn't matter whether you live six feet above sea level or sixty-five, because he, like James Hansen, believes that all of these predictions are, to put it mildly, very, very low. "The rate of sea level rise is currently doubling every seven years, and if it were to continue in this manner, Ponzi scheme style, we would have two hundred five feet of sea level rise by 2095," he says. "And while I don't think we are going to get that much water by the end of the century, I do think we have to take seriously the possibility that we could have something like fifteen feet by then."

It's a little after nine o'clock. Hal's sons stop sipping their lattes and the oceanographic scientist behind me puts down his handful of M&M's. If Hal Wanless is right, every single object I have seen over the past seventy-two hours—the periodic table of elements hanging above his left shoulder, the buffet currently loaded with refreshments, the smoothie stand at my seaside hotel, the beach umbrellas and oxygen bars, the Johnny Rockets and seashell shop, the lecture hall with its hundreds of mostly empty teal swivel chairs—will all be underwater in the not-so-distant future.

One of the few stories I remember from the Bible vividly depicts the natural and social world in crisis. It is the apocalyptic narrative par excellence—Noah's flood. When I look it up again, however, I am surprised to find that it does not start with a rainstorm or an ark, but earlier, with unprecedented population growth and God's scorn. It begins, "When human beings began to increase in number on the earth." I read this line and think about the six thousand inhabitants of south Florida turning into six million in 120 short years. "The LORD saw how great the wickedness of the human race had become." I think about the exponential increase in M&M's, Chobani yogurt cups, and grande lattes consumed over that same span of time. The dizzying supply chains, cheap labor, and indestructible plastic. "So God said to Noah, 'I am going to put an end to all people, for the earth is filled with violence because of them.'" And then the rain began.

I do not believe in a vengeful God—if God exists at all—so I do not think of the flood as punishment for human sin. What interests me most is what happens to the story when I remove it from its religious framework: Noah's flood is one of the most fully developed accounts of environmental change in ancient history. It tries to make sense of a cataclysmic earthbound event that happened long ago, before written language, before the domestication of horses, before the first Egyptian mummies and the rise of civilization in Crete. An event for which the teller clearly held humans responsible.

＊

Dig into geologic history and you discover this: when sea levels have risen in the past, they have usually not done so gradually, but rather in rapid surges, jumping as much as fifty feet over a short three centuries. Scientists call these events "meltwater pulses" because the near-biblical rise in the height of the ocean is directly correlated to the melting of ice and the process of deglaciation, the very events featured in the documentary footage Hal has got running on a screen above his head.

He shows us a clip of the largest glacial calving event ever recorded. It starts with a chunk of ice the size of Miami's tallest building tumbling, head over tail, off the tip of the Greenland Ice Sheet. Then the Southeast Financial Center goes, displaying its cool blue underbelly. It is a coltish thing, smooth and oddly muscular. The ground between the two turns to arctic ice dust and the ocean roils up. Next, chunks of ice the size of the Marquis Residences crash away; then the Wells Fargo Center falls, and with it goes 900 Biscayne Bay. Suddenly everything between the Brickell neighborhood and Park West is gone.

The clip begins again and I watch in awe as a section of the Jakobshavn Glacier half the size of all Miami falls into the sea.

"Greenland is currently calving chunks of ice so massive they produce earthquakes up to six and seven on the Richter scale," Hal says as the city of ice breaks apart. "There was not much noticeable ice melt before the nineties. But now it accelerates every year, exceeding all predictions. It will likely cause a pulse of meltwater into the oceans."

In medicine, a pulse is something regular—a predictable throb of blood through veins, produced by a beating heart. It is so reliable, so steady, so definite that lack of a pulse is sometimes considered synonymous with death. A healthy adult will have a resting heart rate of sixty to one hundred beats per minute, every day, until they don't. But a meltwater pulse is the opposite. It is an anomaly. The exception to the fifteen-thousand-year rule.

From 1900 to 2000 the glacier on the screen retreated

inward eight miles. From 2001 to 2010 it pulled back nine more; over a single decade the Jakobshavn Glacier lost more ice than it had during the previous century. And then there is this film clip, recorded over seventy minutes, in which the glacier retreats a full mile across a calving face three miles wide. "This is why I believe we are witnessing the beginning of the largest meltwater pulse in modern human history," Hal says.

As the ice sheets above Hal's head fall away and the snacks on the buffet disappear, topography is transformed from a backwater physical science into the single most important factor determining the longevity of the Sunshine State. The man seated next to me leans over. "If what he says is even half true," he whispers, "Florida is about to be wiped off the map."

✳

At this point in the story of the flood, God starts giving Noah directions. He says, "Make yourself an ark of cypress wood; make rooms in it and coat it with pitch inside and out."

That evening Suzanne Lettieri, a visiting architectural critic at Cornell, calls me. I am sitting at my seafoam desk, ass plastered to a mod blue stool, transcribing Hal's lecture. She tells me that she studies the impact of sea level rise on architecture; I ask if she has spent time in Miami.

"Nothing substantial yet," she says, "but I plan to this summer." Suzanne reveals that she grew up in Oakwood Beach, Staten Island, a community I have written about. "That's how I found you," she says, "because of your articles." She has seen the impact of sea level rise firsthand. When Hurricane Sandy destroyed much of Oakwood, many residents decided they didn't want to return. They chose to retreat instead. To watch what remained of their homes get bulldozed. To walk away. Suzanne's parents included.

She tells me that she illustrated her latest piece with a collection of photographs she and her husband took on a road trip along the East Coast. "Hold on. I'll send it to you," she says. When I hear "Hold on," I think, with my eyes focused on the world out the window, "To what?" Just beyond the construction crane cantilevered over the base of what will soon be yet another skyscraper, Miami Beach shimmers with its more than 200,000 cubic yards of sand trucked in this year. A century ago this was all one big mangrove swamp. Then a billionaire land developer named Carl Fisher decided to turn it into a resort. First he had the forest cleared with the help of a "machete plow" built and shipped from Indiana; then he dredged mud up from the bottom of Biscayne Bay, pumping it into the acres upon acres where the fallen mangroves once stood. It was an early, labor-intensive form of backfilling, the same process that would shape much of what we consider the Eastern Seaboard today.

Another boldfaced line appears in my inbox. I click on it and open Suzanne's article. Immediately I am drawn to the images that accompany the text. Each one is a black-and-white portrait of a home that has been lifted up off the ground and away from flood risk. They remind me of the work of Bernd and Hilla Becher, a husband-and-wife team who crisscrossed Germany after World War II, photographing the country's disappearing industrial architecture. The Bechers took simple, straightforward shots of water towers, oil refineries, gas tanks, and storage silos. Then they arranged the images in grids, creating typologies that highlighted the formal qualities of each class of industrial structure. Both the contemporary images and the historical ones were taken on cloudy days; the photographers used negative space to frame each silhouette, making the building seem somehow personable, as though it is sitting for a formal portrait. But while the Bechers were showing us the presence of the past, an archive of what once was and what would soon be gone, Suzanne and her husband are showing us the presence of the future, the flood that will be and in many cases already is.

"I've seen houses like these down in Louisiana," I tell her.

"The phenomenon is really wide ranging. Because of Hurricanes Sandy and Katrina, FEMA is redrawing the flood maps," she says, "expanding the area along the East and Gulf Coasts that is considered to be in the high-risk flood zone." To avoid a spike in insurance costs, single-family homeowners in these locations are lifting their houses if they can afford to cover the initial expense. "In some areas the new flood lines go up as much as twenty-five feet, leaving whole neighborhoods perched on stilts, awkwardly looming above the ground."

When I squint, the raised structures in Suzanne's photos look like little boats, lone wooden cabins afloat in a sea of white. Windows are portholes. Railings are gunwales. Dormers and gabled roofs, sails in my imagined wind. A fleet of arks. Some are even fashioned from the cypresses that Louisiana's exploding population clear-cut nearly a century ago. If seas rise six feet, I can imagine these structures floating in place, like ships lashed to their moorings, a stationary flotilla.

Suzanne asks if I know of any areas in Florida that are undergoing systematic managed retreat, neighborhoods she calls "twenty-first-century ghost towns." "We have two choices," she says. "Raise or raze. You either lift your home on stilts to allow the water to move under it, or abandon the structure and move in." She considers Staten Island. "Seeing my childhood home destroyed was an experience," she says. I want to add *unsettling* before the final word of that sentence. I think about how the disappearance of whole neighborhoods is already disarranging our mental maps of who we are and where we come from—and it makes me wonder whether you can unsettle something when the bottom has already dropped out entirely.

I find myself soothing Suzanne over the line. I tell her that raising wasn't an option in Oakwood—the houses were too close together. If a storm wiped one off its stilts, it might have knocked into a neighbor's, and the whole street would have fallen. "Like a set of dominos," I say.

"Can we learn to see demolition, absence itself, as an architectural form?" she asks quietly, before hanging up.

That night I have a strange new dream. In it I am taking a shower in a windowless bathroom when the electricity goes out. The reason for this is unclear, but I imagine it is because of a terrible storm. I grope in the dark to the fuse box and flip all the switches, but the power does not return. I take my laptop into the bathroom and continue to shower by the ambient light the screen throws, all the time worrying about what will happen if the power comes back while I am in the shower. Will the sudden pulse of electricity somehow electrocute me?

The light does not return and I am not seized in its current. But my computer screen eventually dims and the water in the shower cools. I sense the world beyond my windowless wall is changing, but it is hard to tell how much when I am having such difficulty seeing.

*

In 1850 the Swamp Land Act passed through Congress, granting states the right to sell federal wetlands to individuals. The resulting funds were earmarked for drainage and levee building, costly interventions in the landscape that would, it was hoped, help convert areas previously "unfit for cultivation" into a vast network of agriculturally productive farms. A speculative frenzy began, reshaping wetlands—both tidal and fresh—across the country. But perhaps no place was as profoundly transformed as the newly minted state of Florida, where twenty-two million acres of marsh (or 59 percent of the state's total landmass) were handed over to developers.

In the intervening century and a half, the unique watershed that covers south Florida in wetland ecosystems of all sorts— from cypress stands to saw grass marsh, from mangrove swamps to peat bogs—has been reduced to less than half its former size. As Susan Cerulean writes in *The Book of the Everglades*, "Sixteen hundred miles of canals, levees, and pumping stations, the most extensive water control project in the United States, have bent and restrained this watery wilderness into a more predictable, flood-free landscape." Land one US Army officer had just over a century ago referred to as "swampy, low, excessively hot, sickly and repulsive in all its features" was drained and refashioned as a kind of neotropical paradise that supports multiple billion-dollar industries—citrus, tourism, and retirement among them—if not much else.

What remains of Florida's wetlands—many of which run right up against Miami-Dade County's western border—is now being threatened from the other side. As sea levels rise, the interface between salt water and freshwater is pushing deeper and deeper into this fragile ecosystem, transforming the marsh from within. The saw grass found in the Florida Everglades is somewhat salt tolerant but, like many marsh grasses, has a threshold. When ex-posed to a salinity level beyond that threshold, it puts fewer roots into the soil, microbial activity kicks up, and the ground around the plant slumps and loses elevation.

"The formal term for the phenomenon is 'peat collapse' and it isn't only happening here. There's evidence of it in Louisiana, and all the way up the East Coast," says Tiffany Troxler, the lead researcher in a National Science Foundation–funded study on drivers of the condition. "In the context of trying to set up this experiment, we learned that brackish waters were much more prevalent in the peat marshes of the Everglades National Park— were much farther inland—than we had anticipated."

I spend the morning after my conversation with Suzanne Lettieri waist deep in marsh muck, in the heart of Florida's signature national park. Members of Tiffany's research team are pumping salt water into test plots of peat, taking salinity readings and root bundle observations. We are twenty miles inland from the Florida Bay, just beyond the northernmost example of peat collapse. The scientists are trying to simulate the process to better understand what exactly is causing the marsh to deteriorate. "We brought the assistant science director of the park out because we wanted to show her the condition of the site. It was the first time she had ever actually seen evidence of peat collapse in the Everglades. Up until that time she'd experienced it only through landscape photography," Tiffany tells me. "When she actually got out of the car and into the marsh she was stunned. I can still hear her saying, 'Oh my god, what is going on?'"

My second-grade science fair project, a haphazard investigation of Florida's topography, was called *How Flat Can You Get?* I painted a canary-yellow sun on the poster board and illustrated my findings with a hand-drawn map, above which I listed my favorite factoids. "Florida is the flattest state in the union!" I wrote in shaky purple script. "Did you know present-day Florida used to be under the ocean!" And: "The highest point in Florida is 345 feet above sea level, which is ¼ the height of the Empire State Building!" I did not write, "In Greenland chunks of ice three times bigger than the Empire State Building are crashing into the sea, slowly drowning present-day Florida," because twenty-five years ago the ice wasn't calving and Florida wasn't drowning. My project didn't win a prize; it wasn't even given honorable mention. Back then topography probably seemed boring, especially when set next to dioramas of rain forests and exploding miniature volcanoes.

But today topography is perhaps the single most important element contributing to Florida's profoundly precarious position. Southern Florida is so flat, so close to sea level, that a three-centimeter jump in the height of the ocean, as has occurred over the past decade, has already dramatically affected not only seaside human communities but the flora and fauna that have long called this place home. Mangroves are migrating in, and many of the species that compose the area's signature hardwood hammocks—Jamaican dogwood trees, stoppers, pigeon plums, buttonwoods—are all starting to brown and die. These trees like a mix of fresh and brackish water, and the amount of saline in the aquifer they tap is on the rise.

After finishing up at the test site, I drive the twenty miles down State Road 9336 to get a good look at the bay that has begun to work its way into the Everglades. At first glance it appears utterly harmless—boring, even. Just another body of water the color of concrete. I eat a Cuban sandwich and call Pete Frezza, the research manager for the local branch of the National Audubon Society. When I ask him how sea level rise is affecting the species he studies, he doesn't miss a beat. "It used to be that there were over a thousand roseate spoonbill nests out on the islands of the Florida Bay—Sandy Key, Tern Key, Frank Key. Those were *the* spots," he says, as though describing a secret surf break. "But now we can't find a single nest out there."

Spoonbills, Pete says, are specialist eaters. In this they differ from other wading birds. Herons and egrets, for example, feed in a variety of places, with a variety of fish densities; if a lone fish swims by, they can see it and go after it. The gangly pink shore-birds with beaks the shape of serving utensils, however, require extremely high concentrations of fish just below the surface. That's because they don't use their eyes. Instead they walk around, swinging their bills back and forth, blindly praying for fish.

If spoonbills can't find adequately shallow water, they can't feed, and if they can't feed themselves they can't feed their chicks. "During nesting season, spoonbills require a water depth of only three or four inches, five at the most," Pete says. "If the water is too high on the spoonbill's foraging grounds then their nests fail and their chicks die." From my rental car, the gray bay now seems less benign. "The spoonbills are in big trouble. The kind of environment they need, and it is a very particular environment, has mostly disappeared out here. Over the past ten years we have witnessed an unexpected and unprecedented shift of every single spoonbill nest north, onto higher ground."

It used to be if Pete saw a spoonbill he knew he would also find snook, redfish, and kingfishers. That's because spoonbills feed on the same small fish the big fish eat. But the addition of just three centimeters of salt water in southern Florida's wetlands has significantly changed the landscape over tens of thousands of acres of marsh, eliminating the foraging habitats of these birds and causing widespread peat collapse and telltale hardwood death. This is affecting animals throughout the whole trophic structure, from microorganisms all the way up to megafauna.

"If you don't have spoonbills, you're not likely to find the game fish either. For myself, and for my buddies who work as guides out on the Keys, that is an important shift. This friend of mine named Dave, he talks a lot about moving inland. His income is disap-pearing, the housing cost keeps going up, flood insurance is on the rise, and you add the disappearance of the fish he used to rely on to pay his bills. It's starting to seem like throwing good work behind bad for him to try to stay." Dave had begun to understand that his way of life was in danger. The attention he paid to the land, and its changes, had made him think that perhaps the time had come to leave.

"I mean, look at the spoonbills," Pete says. "They appear to be abandoning Florida Bay. It's as simple as that." Then he hangs up.

*

The following Friday, as I drive from my hotel back to the Cox Science Center, I think about what would happen if all the water stored in Antarctica's and Greenland's ice sheets were released. The stilts in Suzanne's photos would anchor those homes in place while the Atlantic rose up, up over the driveways, up over the to-piaries and the stairs, up over the decks and the railings, up over the windows and the roof gables, until the homes disappeared beneath the flat blue surface of the sea, where clouds would skate and bloom as though nothing and no one had ever existed below.

I drive past the high-rises currently under construction, with breezy names like Aria on the Bay, One Paraiso, and Solitair. Past two Lamborghinis, two Ferraris, one Rolls-Royce, and one brand-new Bentley with a matte white coat of paint and chrome hubcaps. Past the port where six cruise ships pause. Past a string of coin-size islands covered in not-so-coin-size houses: Star and Fisher, Belle and Hibiscus, San Marco and Rivo Alto. And as I pass, I imagine all of it underwater.

Past the new Pérez Art Museum, which already sits on fif-teen-foot stilts as a safeguard against higher tides, stronger storms, and severe flooding. Past the Crescent Heights Inspirational Living construction site on the corner of Alton Road and Sixth Street—the intersection that flooded twice in 2000, four times in 2010, and eight times in 2013. Past the floodwater pumps and the street-elevation projects meant to serve the imagined residents of the "Wave," the name Crescent Heights Inspirational Living gave their new development. Past CVS, Walgreens, H&M, and Forever 21. Past Petco Animal Supplies and Wet Willie's cocktail lounge. Past the South Seas Hotel and Eden Roc, the Shore Club and the Ritz. Past a couple with sun hats and beach chairs and a dachshund.

In my mind there is no tidal wave and no wreckage. Instead everything is simply, coolly covered by blue.

I am ashamed to say that when I finally reach the Cox Science Building, I sit in my rented Toyota Yaris and feel something close to smug. I imagine this is how the Oracle of Delphi felt divining the future from a fistful of smoke. "Love of money, nothing else, will ruin Sparta," she said to Lycurgus. Yes, I think, love of money, nothing else, will ruin Miami.

But then I remember that water is not discerning. It doesn't know the difference between a spoonbill and a skyscraper, between a millionaire and the person who repairs the millionaire's yacht. The thought stops me in my self-righteous tracks.

✳

A few days earlier I had spent the afternoon in Shorecrest a neighborhood a couple of miles north of downtown. To get there I leave the beach behind and drive past Arky's Live Bait & Tackle, Deal and Discounts II, Rafiul Food Store, Royal Budget Inn, Family Dollar, and Goodwill. As I continue north, the buildings all lose their mirrored glass and their extra floors, until most are single story and made from stucco.

It isn't raining when I arrive in Shorecrest, and there isn't a storm offshore; the day is as clear and as blue as the filigree on a porcelain plate. But the streets are still full of water. I watch as a woman wades ankle deep across Tenth Avenue. She has gathered her long russet-colored skirt in her right hand, and in her left she holds a pair of Jesus sandals. When she reaches the bus stop, she sits and puts her shoes on. On the corner a man stands facing traffic, holding a sign that simply reads, "Please help FOOD."

"We get flooded with just about every high tide," the woman tells me, glancing in the direction of the panhandler and the approaching bus. "And if the moon is big it's worse."

All along the East Coast, from Portland, Maine, to Key West, "sunny day flooding" is increasingly frequent. Many places in the Sunshine State are so low lying that high tide—when coupled with something as innocuous as a full moon—can cause the streets to brim with water. Sometimes the tide simply rises above the seawalls and starts to spill into the roadways; in other cases it enters the neighborhood through the storm-water infrastructure belowground. The very pipes designed to reduce flooding by ushering rain out instead give salt water a chance to work its way in.

In Shorecrest I spend a minute watching the bay burble up through the street grate and onto Northeast Little River Drive before whipping out my camera and snapping half a dozen photos.

Just then a man walks up behind me, peers down, and says, "I've seen fish come swimming out."

"No, you haven't!"

"I have," he says, pushing his sunglasses up. "I've been here twenty years. When I first moved we used to flood once a year, maybe twice. Now it's constant." His name is Robert Cisneros. He grew up in Cuba and moved to Florida in 1962, dropping the final *o* in Roberto. He thought the name change might help his fledgling boat repair company succeed.

I had heard about the flooding in Shorecrest from Nicole Hernandez Hammer, a sea level rise researcher turned advocacy coordinator for the Union of Concerned Scientists. The daughter of Cuban and Guatemalan immigrants, Nicole left academia for activism when she observed firsthand the disproportionate risk that climate change poses for the Latinx communities of south Florida. "We know that people of color are often the most vulnerable to climate change, and that these communities also tend to receive disproportionately low funding for adaptation, resiliency, and relocation," she told me over the phone. "You're staying out on the beach, right? Just drive up to Shorecrest. It will take you twenty minutes, and you'll see what I mean."

Robert points to his house and his yard, which are catty-corner from the drain we stand by, and says, "I used to have a nice garden here, and now you see how it is. The water comes in and sits. And everything dies because of the salt. It's not rain that floods this place. It's the ocean. I just bought some stones to put here to try to keep the water out. But other than that, what can I do?"

I ask if the city is helping the neighborhood come up with short-term solutions. Robert gets upset. "I think they need to raise the street. They need to install pumps. But those kinds of things only happen on the beach. They're not giving any of us here any relief."

Like Miami Beach, Shorecrest was built atop a former wetland. On the strip, where billions of dollars in real estate investment are at risk, the government is using a mixture of property taxes and municipal bonds to invest in formal sea level rise adaptation. But in Shorecrest, Hialeah, and Sweetwater—low-to-middle-income neighborhoods where the majority of residents are people of color and municipal services have long been difficult to maintain, thanks to the discriminatory banking practice known as redlining and the resulting decline in property taxes—residents are expected to remove their shoes and wade through the water.

Robert shakes his head in disbelief. "I wanted to leave this house to my kids, but soon it's going to be worthless," he says. On his stoop sit two pairs of rubber boots, ready for the flood that is already here.

❋

Back in the parking lot of Cox, the past decade of storms goes thundering through my mind. I see large waves pulsing into the Lower Ninth Ward. Roiling chocolate milk ripping homes off their foundations in Oakwood Beach. Family cottages in the Rockaways turned into aquariums. And the water that shut down the elevators in the Red Hook Houses for more than a week, stranding the elderly unable to descend multiple flights of stairs.

I remember walking through the Lower Ninth five years after Katrina. The storm's eerie language was still clearly legible on one house's laminate siding: a painted X, its upper quadrant marked to show one dead in the attic. On the next building, a lighter, more haunting X showed two dead on the ground level. Someone had attempted to scrub it away, to remove the memory of the killing wind, the way the water deformed the foundation and window frames, the lives it took.

I turn the key and take it from the ignition. Lock the doors behind me. Outside, cicadas rattle the humidity.

My thoughts are tumbling one over another.

Wendell Berry writes, "It is the destruction of the world in our own lives that drives us half insane, and more than half." The world isn't only the physical universe of objects outside the body; it also hums within the mind, is the constellation of thoughts we have about tangible matter. Destruction lingers, takes place many times over—once in the moment of violent dissolution and also much earlier, when we learn to think of this derangement as possible. When we learn to acknowledge that the water will come. Then just imagining an end to the world as we know it means also, at least partially, losing your own mind.

I start descending the poorly lit staircase to the basement of Cox, where the geology department is—fittingly—located. On my way I pass seven blown-up photographs of students working in the field. A group stands beside Mount Saint Helens, a couple of weeks after the volcano exploded for the first time in 127 years. In another, fresh-faced undergrads don waders. They sink waist deep in mud in order to collect a core sample from a soggy marsh. There are six-foot-long maps of the tectonic plates running beneath the United States, and also photos of students snorkeling in coral reefs. I smile back at their grinning faces, but it feels forced.

The atmosphere in Hal's office is more in keeping with my mood. He has come down with a cold and is just back from a trip to Publix, where he bought a bunch of fruit platters for another climate change conference. He is wearing the same taupe suit as the week before. The skin below his eyes is the color of charcoal, and the rest of his face sags too. Still he manages a smile. Up close it is hard to imagine that the media has nicknamed this good-natured septuagenarian "Dr. Doom."

Hal pushes some of his scattered papers aside so I can set up my recording device. A woman in a quick-dry army-green dress offers me a cup of coffee. I have my question, just one, prepared: "What single event woke you up to the reality of sea level rise?"

In response, Hal walks me through half a century of climate science breakthroughs. An early report by a researcher named K. O. Emory at Woods Hole, who noted that the levels of the Atlantic were changing. Michael E. Mann's climate modeling that suggested humans have been affecting the temperature of the earth's atmosphere since 1950. The discovery that most of the warming from greenhouse gases is actually sinking into the sea and causing its water molecules to expand.

I write all this down even though I know it already. Even though it has nothing to do with Hal specifically. I ask my question again.

"You know, twenty years ago I never thought I would end up seeing the rise because everything, all the projections at that time, really didn't ramp up until well into the twenty-first century. But then I started going out to Cape Sable." Cape Sable is the southernmost part of the mainland; it reaches into the Florida Bay like a swollen hook. "Out there the beaches were disappearing, mangroves were moving in, tiny channels turned into huge rivers in a matter of years. Even the roseate spoonbills started abandoning their nesting grounds. I had never, in my life of studying the geology of the coast of Florida, seen anything like it. That is when I knew in my gut that the early predictions were wrong and that sea level rise was unfolding a lot faster than any of us ever imagined."

Neither of us says anything for a while. Hal taps his pen on the desk and tells me he has only ten more minutes to talk.

"What comes next?" I ask.

What I really mean is: Who will get to enter the boat? When God says to Noah, "Go into the ark, you and your whole family.... Take with you seven pairs of every kind of clean animal, and one pair of every kind of unclean animal, and also seven pairs of every kind of bird, to keep their various kinds alive throughout the earth"—I think perhaps he is also speaking to Hal Wanless. I want to say: If I believe you, will I be among the chosen? I want to say: Will those who are the most vulnerable now have the same chances as those who are not?

Hal tells me about his pairs of clean and unclean animals. "We have to start relocating the things we value," he says.

"Like the Smithsonian Institution, which is sited on top of an old marsh. We have to make seed banks, a global archive for the future, and we have to move our power plants, in order to maintain a functioning society. We have to start lining the trash dumps that line our shores, we have to start preparing for inundation. Remember, the last time carbon dioxide levels were the same as they are today, the ocean was one hundred feet higher."

The last time carbon dioxide levels were this high was during the Pliocene epoch, 2.6 to 5.3 million years ago, when megatoothed sharks prowled the oceans. The last time carbon dioxide levels were this high, California's Sierra Nevada rose up and tilted its granite face west. Shifting plate tectonics re-opened the Strait of Gibraltar, forming the Mediterranean Sea. The last time carbon dioxide levels were this high, armadillos migrated north across a newly formed land bridge between today's North and South America. Dogs headed in the opposite direction. But no one can remember these things, because humans didn't exist.

What we might remember, however, if only faintly, is this: Fifteen thousand years ago, carbon dioxide levels increased rapidly and so did the height of the sea. Fifteen thousand years ago, human beings were transforming from hunter-gatherers to farmers. Fifteen thousand years ago, we domesticated the first pig. Roughly fifteen thousand years ago, the woolly mammoth started to go locally extinct and with it went the giant sloth.

And fifteen thousand years ago, meltwater pulse 1A occurred. Hal points to it when people ask if sea levels have ever risen rapidly before. Some even speculate that meltwater pulse 1A might be why we have the story of the flood. A story where "Noah and his sons and his wife and his sons' wives entered the ark to escape the waters.... Pairs of clean and unclean animals, of birds and of all creatures that move along the ground, male and female, came to Noah and entered the ark."

Back when meltwater pulse 1A was happening, human beings did not have written language. That would come ten thousand years later. If they were among those who walked onto the ark, they could tell their children about the world before the water, about the killing wind. And their children could tell their children, and a ten-thousand-year-long game of telephone would begin, in which, of course, some of the details would get scrambled.

Was there only one giant boat, or a flotilla of tiny ships?

Did Noah wear a tunic or a taupe suit?

Was it for forty days or four hundred years that the waters rose?

I once heard a poet read from a book she wrote after Katrina. She said—in a voice I understood as God's voice, on the night the water arrived in the city—this:

What do you expect me to do

I am not human

I gave you each other
so save each other.

I leave Hal's office and drive back across the bay. That night the moon is big and full, making the ocean weightier. The salt water unspools in the streets and continues drowning, by degrees, the low-lying land that lines our shore. Soon, I think, if Hal is right, all of this will be underwater, not just temporarily, but for good. In the meantime, in Miami Beach the water pumps whir, while five miles north of here there is a barefoot woman who carries her sandals in her hands, wades through the all of it that is always now rising.

On Reckoning

Dan Kipnis: Miami Beach, Florida

TODAY I AM PACKING UP MY OFFICE, WITH ALL MY MEMO-
rabilia—my world's records and mounted fish, things I have collected over the years. Now that my possessions are boxed, the
walls are bare. All of my International Game Fish Association records have been beaten except for one—the black arowana. I actually caught it in the Río Bita, a tributary of the Orinoco, on the
Colombian side. It is a beautiful river. The fishing is outstanding.
You can go up these little tributaries and there you find peacock
bass, arowanas, giant catfish, prehistoric-looking stuff. I made a
number of my records way back in there. So it is hard to see all
this stuff packed away.

My wife has been in this house thirty-eight years. It was built
in 1956 originally; a small home, right in the middle of Royal
Palm Avenue. So my wife and I said about ten years ago, "Let's
remodel." We knocked down everything except the front wall.

I had a contractor come in and shell the place out. Then I spent a year working. I've got a lot of sweat and blood in this place. My wife was very kind, very kind, to let me do it. We ended up with a hellacious house. The whole downstairs is open. The kitchen, dining room, and living room are all one room. It is absolutely beautiful. I have two bedrooms downstairs, on the front of the house, as it was in the original. Then you go upstairs and there are two bedrooms and the whole master suite. Oh, it is gorgeous, a big curved wall behind the bed. They wanted to lower the ceilings but I told them absolutely not, I just followed the roofline.

We never thought we would leave here. This was the house we were going to grow old in; we were going to love each other until the end in *our* house. Then it would go to the kids. But I'm not going to be able to give the house to them because of what is happening in Miami Beach. Because of sea level rise.

Nobody in the government wants to discuss it. People think if you move west you are safe, but that was all the Everglades. Let's assume that we get about six and a half feet of rise by the end of the century, which, according to most scientists, is a pretty conservative estimate. If we get six and a half feet, there is very little land left in south Florida, and what is left is kind of like an archipelago—surrounded by rivers and transverse glades. Everything is marshy and there are no roads. The west coast is gone. The beach is gone. The east side of Biscayne Bay is gone. The infrastructure is gone. So I just don't see how people can live here with six feet of sea level rise. When I hear the mayor of Miami Beach say, "We will be here for the next two or three hundred years, we are going to design a way out, technology will save us," I think, *Bullshit.*

You know what makes life so great down here? Our beautiful weather—and you know what, it's going to be hot as hell. We won't have any freshwater. We are going to have mosquitoes all over the place. We are going to have diseases vector in, yellow fever and dengue. And besides, everything is going to be underwater rotting.

All the streets. All the trees. All the septic tanks. All the roadways. All the vegetation that has died because it can't survive all the salt. How about that? Doesn't sound like a very nice place to live, does it?

So, you know, in a sense we are very lucky. We are being forced to move and yet because we are leaving now, before people really understand the extent of what is to come, we have the ability to make a handsome profit, which will afford us the ability to live wherever we want. We can't replicate this place, but we certainly can move somewhere nice.

But what happens if you're a poor person renting in Miami? No one is going to help you move. No government is going to chip in. Or what if you own your house and you have a mortgage on it and you can't get insurance and no one wants to buy it because it floods? You lose your whole equity. In Miami-Dade County we have almost three million residents, and 60 percent of those households don't make a living wage. Many live with two or three families in a single residence. They are working hard trying to get a better life, trying to get the American Dream. Well, you know what, their dream is gonna drown.

My time frame on this is way different than most. I have a pretty good idea of what's coming and what every six inches of rise will do. The Miami Beach Marine and Waterfront Protection Authority has tackled causes and issues that were never within the scope of the organization before I became chairman five years ago. Now we are trying to look at things through the eyes of a city that will be inundated. We are experimenting with living seawalls and raising roads. And these interventions will buy time for us to get the hell out without having to run away; they will help us make the transition without as much pain and suffering as we could otherwise have.

What's not easy is giving up something you've been totally invested in your whole life, and for me that is Miami Beach, not just the house but the community. In the 1920s my father's father built a place on Palm Island, just off the MacArthur Causeway. I grew up in that house, on that island in the bay. I have been on the water my whole life. I went to high school by boat. I had a thirteen-foot Boston Whaler that I would drive right up the Collins Canal to the front of Miami Beach Senior High. Back then there was a bridge going across the canal at Washington Avenue. I would take the gas line off so no one would steal the boat, tie up there, and go to school.

What I am trying to say is I belong here. I go to the grocery store and I know all the people, even though it is one of those big-box Publix. I know the butchers and the guys stocking the shelves, and I know the produce guys and the managers. I know my postman. I know my garbage man. You know what I mean? It's a small town. I know the mayor. I know the people here, and I have worked my ass off trying to make this place better. And it's hard to walk away.

But I'm tired of this fight. I will still lecture to young people 'cause that is where the future is, OK? But I'm done with the older people. I'm done with the governments. I'm going to buy another house where it's high and I don't have to worry about sea level rise and then I want to go to Australia and New Zealand, the South Pacific before it goes underwater. I want to go around South America again, I want to go to coral reefs; I want to go to Spain and Portugal. I want to go to Italy, drink wine, and eat good food. I want to go fishing where I like and let them fight it out.

When I finish my talk to high school and college kids, I say this: You're standing there and the government is twenty feet away from you and the government has this giant revolver with a bullet in it that is huge, it's just huge. You're twenty feet away and you're staring down the barrel of that revolver and the government decides to pull the trigger and you see the smoke come out and there is a flash and the bullet, it's coming right at your head in slow motion. So what are you gonna do? It's coming real slow but if it hits you, you're dead. Are you gonna stand there and let it hit you right between the eyes, or are you gonna duck?

PART II

Rhizomes

On Storms

Nicole Montalto: Staten Island, New York

I WAS LIVING AT HOME AT THE TIME OF THE STORM. I WAS working at a dentist's office and living at home. Of course I had my own bills, like my cell phone bill and my car insurance, but I didn't have a mortgage and I didn't have to pay utilities. I stayed with my father because I needed to save money. He was never the type of guy who would call out for help. If something needed to be done, we just did it. We figured it out.

My father lived in that house since he was six. He bought it from his father in 1984, around the same time he married my mother. When I was little there weren't as many houses and the wetlands were larger. We all had the same bus stop; we all went to the same school. Me and my sister were big into playing soccer and kids from the neighborhood would come over 'cause ours was one of the bigger yards.

Even as we got older, if my friends got in trouble and needed a place to crash, they came to us. Sometimes my dad would bust our chops about things, like, "All right, your friend has been sleeping on the couch for three days now, what's going on?" But mostly he was cool. Katie—she's my best friend to this day—she would call sometimes just to speak with my dad. He would shoo me away from the phone and I could hear her saying something like, "My brakes aren't working right," and my dad would tell her to bring her car over, that he would fix it for her.

I am twenty-six now; I was twenty-four at the time. The year before, with Irene, the press and everyone made it up to be a huge thing. My sisters and I, we brought the animals to my mother's apartment in Annadale. I had a dog that was so old—he was blind and deaf—that I felt bad moving him. But Irene was nothing, nothing happened. With Sandy they were making an equally big deal, and you know, some people were taking it seriously, some people weren't. My dad being one of those people. We thought it would be the same thing as the year before.

I'm not really an anxious person, but for some reason I was fidgety the day before the storm. I even packed a bag, just in case we had to get out quick. I put clothes inside ziplocks, I put my electronics in there in case I was bored, and I packed another bag for the animals. I didn't sleep much.

Around ten o'clock that Monday morning, I called my dad because the backyard started to flood. He decided to come home early from work. Flooding wasn't the most uncommon thing. There was a certain part of my yard that got wet—but it had to be a lot of rain.

He came home and me and him started doing rundowns. We went through the garage and basement, asking, *Is there anything in here that we should bring up higher, just in case?* My older sister's room was in the basement, so we put some stuff up on shelves. I remember moving her clothes, electronics, things like that. The rest of the rooms were upstairs. That was where the things of real value were. We brought the cat litter up because that's something you really don't want floating around.

My little sister is a bit of a nitpicker. I was like, "You know what, let me bring you to your friend's house." If something was to go down, me and my dad would be OK. Whereas with my little sister I felt like we would have had to smack her, put her in the car, and go. She had a friend, Frankie, that was home, and he was up high on a hill. I drove her there at 1:00 p.m. My older sister was at work or at her boyfriend's or something. So it was just me and my dad the rest of the day. It was like a regular day. Actually, because I hadn't slept the night before, I took a nap.

I woke up and the storm had started. It was getting dark. My father was going to run for president of the postal union that year. And that was what he was doing; he was on the computer typing up the speech. There was one small desk lamp on. I remember a tree falling on the other side of the garage. We watched it and were totally nonchalant about the whole thing.

My neighbor across the street, my aunt Patti's daughter's husband, came running over around seven. "It's flooding fast, it's up to my ankles now," he said. I really didn't plan on leaving, but there was this panic in his voice. That was when my dad was like, "You gotta go." What we had done, actually, was I had parked my car up where I thought it was safe from the water, about a mile away. My backup plan was to go to my mom's. My dad didn't really get along with my mom's sister, so his whole thing was, "I'm not leaving." He said, "You know, I just want to make sure the pump is working." That the electricity was off. All those things you think about when it floods.

At that point he was like, "You gotta go, you gotta get outta here." The assertiveness in his voice told me it was a no-joke sort of thing. He said, "Take my car; go to your mom's." I was worried about the animals. But my father said he'd take care of them. By the time I left, I had just enough time to get the car out, because on Mill Road it was deeper than at my house by a couple feet at least.

I went out the one way. My dad yelled out the door, he was like . . . [*Interrupts herself.*] I don't like to talk about this stuff, I rarely ever do. [*Crying.*] Sorry. This is why I don't do interviews. But you're writing a book, and that's different because it will immortalize him.

He yelled out the door, "Don't hit the brake, just slam on the gas!" So that's what I did, I just took off. If he hadn't said that, I might have hit the brakes and gotten stuck five seconds from my house.

I slammed on the gas and went flying through the water.

There was a huge tree down on Guyon Avenue. I had to take side streets just to go around it. I got to my car and I called him. To say I made it, I'm driving now. I told him there was a tree down but he sounded preoccupied. He said it was a good thing I got out when I did. "The water is rushing in," he said. "I gotta go, I gotta go," and he hung up.

I got to my aunt's. If I had to guess the time, I probably left around 7:30, and that was when the water was suddenly getting pretty deep. I probably talked with him at 7:35 and he said the water was coming in the basement. Maybe I got to my mom's at 7:50. I had only been there for twenty minutes or so. My sisters started calling me, and they kept saying, "I haven't been able to get in touch with Daddy! I haven't been able to get in touch with Daddy!" And I was like, "OK, let me call him." I had known that he was doing something at home, so I didn't want to call him right away. Then I tried calling him and calling him. I started texting him. "Just let us know you're OK."

There was no response. I told them that if by 8:30 he doesn't get in touch I would go back. When 8:30 came around, the storm was really bad. So I decided to give it another half an hour. But at a quarter to nine I had to go. My mom said she would come too. The weather was crazy. There were trees down everywhere. I couldn't even get to Hylan Boulevard because there was so much water. I had Nike thermal pants and all of these tight thermal clothes, boots, everything. I knew I was going to have to tread through water. I was prepared. So I told my mom to wait in the car, I was going to see how far I could get before the water got really high. I didn't even get to Hylan Boulevard before the water was at my waist. I was still well over a mile from my house. I can't swim a mile. If it was at my waist and I didn't even cross Hylan, I couldn't imagine how deep it was over there.

We ran into this cop and he said that T-Mobile was down, that there were power outages everywhere. I was like, "Oh, that makes sense as to why my dad didn't answer his cell phone. OK, thank God, maybe he's all right." The cop told me that there was a back way that I could get closer to my house, down Tysens Lane. It was clear. They had electricity over there. It was like a whole other world. I went down Tysens, I went down Falcon, and I parked as close as I could. It was 9:30 or 10:00 and the water had started to recede a bit. There was a bunch of fire trucks and cops. They didn't know what to do. Nobody expected it to be as bad as it was. As I was walking by, nobody said anything to me. They didn't care that I was walking into the water.

I got to the point where the tree was down, so I was getting closer to my house. There were parts where I was on flat ground, parts where the water was up to my knees and my waist, then flat again. But when I got to the tree, you could see that the water was deep over there. I had all intentions of just going.

In the first flood [the Nor'easter of 1992] my uncle Charlie got me out of my house by putting me on his shoulders when the water was up to his chest. One of the things he had said is that you can't walk in the middle of the street because the manhole covers get blown off during the flood, and then you can get sucked down into those open holes as the floodwaters recede. I didn't think about it at that point. I would have walked down the middle of the street, not thinking about it at all. But there were three guys about to go and then suddenly they were like, "Oh no, forget it." It made an impression. One of them had said, "No, you'll get killed, you'll get killed. The manhole covers." I remembered my uncle Charlie then and I didn't go. I just stood there.

You would see random people running out of the water, going different ways. I kept asking if there was any way to get to Fox Beach Avenue and everybody said there was no way.

I had to turn around and go back to my mom. I told her that it was beyond my capabilities. It's one thing to swim a little and another to swim over half a mile. I'm not sure if you know my uncle's story. He stayed on a neighbor's roof for four hours. You know, in a situation like that, when you have no options—maybe. But why would I walk into it?

It had stopped raining at that point. I walked back to my car and drove to my mom's. My mom and I slept in the same bed, but all I could do was worry. The next day at 7:00 a.m., I was like, "OK, the sun is up, we gotta go." And we did, we went out there. It was insane. Everything was turned over. The watermarks on the homes were so high.

I went into my house; I was screaming for my dad. Everything was upside down. The couches floated to different areas, my bed was up on the wall. The only things that didn't move were my dining room table and the filing cabinet, because both were too heavy. That was where my dog took sanctuary, on the filing cabinet. My cat was sitting there on top of my bed. I didn't see my dad. I thought, "Shit, maybe he left. Maybe he went to someone's house." But then I thought, "He wouldn't leave the animals."

At that point I could actually see the water in the basement. It was still so high. There was maybe at least four and a half feet down there. I was yelling for him. But the only thing I could do was get the animals out. I threw the cat into the carrier and I took the dog under my arm. I went to my mom and told her I couldn't find him.

Nothing was working. From that point on, we went crazy trying to find my dad. Everyone was calling hospitals and precincts. That was Tuesday. We thought maybe he hit his head and forgot who he was. Maybe he was in a hospital as a John Doe. It started to get really cold. We thought, you know, god forbid he was wet, had hypothermia, got dragged away into the weeds.

My dad's wallet was still in his room. There was something else too that he left behind; I don't remember what it was. The wallet was the biggest thing—it had money and his ID. He wouldn't have left the house without those things. Oh, I know what it was. He was on and off with smoking cigarettes—I'm the same way too. And he had left his pack at the table and a cup with a couple of butts though we never smoked in the house. It is so weird to timeline things. I spoke with him on the phone. He said it was good I got out when I did. The water was rushing in. He was on the basement stairs when he was on the phone with me. Did he come back up, and smoke two cigarettes? Was that before or after? I saw these things that were clueing me in to the idea that he never left the house. When I started seeing those things, I went down to the basement and began screaming. I was hoping that I would hear him but at the same time I wasn't.

The walls in the basement were all down. Everything was everywhere. There was a washing machine in front of my face, I could see into my sister's room. There was no way I could walk through. Everything was in the water, floating. I believe that the water came in one big force. [*Silence.*] It's the only thing that makes sense to me.

My dad's friends, once they knew he was missing, they broke all the windows in the basement to get the water out. They started pumping the water too. People say he was down there for the pump, but I don't see how it could have been the pump, because the basement was already flooded. What the hell could a pump do? When we pulled up on Wednesday morning, my father's friend told us that they'd found him. My father was in my sister's room in the basement.

My mom had to go identify the body; even though they were divorced they remained close. She said he had a gash on his head, so what we believe, or what we hope happened, is that he was knocked unconscious. A force of water must have knocked a piece of furniture on him, or knocked him off his feet.

My boyfriend and I had planned on moving into the basement, to live there together, renovate it, and save up money so we could buy our own home. We didn't look for apartments right away because we had to plan the funeral. By the time we got around to it everything was rented out. So we talked my mom into getting an apartment with us, and a year later we ended up buying a two-family home in Stapleton Heights. It's still my mom, my sister, my boyfriend, and I, but we have separate doors now.

After the storm we were all like, "We're moving to a hill," and I moved to a hill. By the time I was twenty-six, I lived through two floods, one of which took my father's life.

I hate when people write comments like, "Well, you shouldn't have lived there in the first place." Of course if we knew, we wouldn't have been there. People don't move into these places thinking, "Living here I might lose my life." No, there are builders who buy the lots and then they sell them and they spin it and you think you are living in a fine house. People buy what they can afford. The problem is that the builders build these homes and the city allows it just to make some money. They took a spot where the water used to be absorbed and they paved it. What they should have done is left it alone. I mean we were right in the middle of a wetlands. And then for people to condemn us for buying the homes, they need to get a life and shut up. That pisses me off more than anything. That's why I don't like doing these interviews: because it puts me out there and it puts my family out there. I have heard people say comments with my father's name in their mouths. I'm not going to get into an argument online, but I see these comments and it hurts.

It's tough to see this neighborhood that I grew up in, that my father grew up in, that my sisters grew up in—I mean, we spent our entire lives there—being demolished. But on the other side, it's nice knowing that this is to protect everyone else and that it can't happen again. At least it can't happen to the people I know and the people I love. And maybe the government really will do the right thing and let Oakwood go back to nature.

Home was that house—it was my dad, it was my mom, it was my sisters. When my dad was gone, it wasn't home anymore.

We'll hang out here at my aunt Patti's for a little while longer. [It is the second anniversary of her father's death.] I'll eat lunch, go home, change, put something in the slow cooker. But the plan is to go to my old house later, one last time before they tear it down. My boyfriend and I, we spent Sunday afternoon cutting down the weeds over there. Someone wrote an article about how they are boarding up all the homes in Oakwood and demolishing them and my house is the first picture. I was so embarrassed, because our house looks terrible. Last year we went there as well, we went there and had some beers and we got together to celebrate his life.

My dad used to play guitar and someone had a disc of him playing. My little sister and I haven't looked at the home movies yet.

We haven't been able to watch them, but we dried them in rice. So that night—a year ago today, actually—was the first time we heard my father's voice, and we cried but it was nice, it was really nice. He was singing and playing the song he always used to sing: "Wild Horses."

Divining Rod

Staten Island, New York

THIS IS A BOOK WITH MANY BEGINNINGS. ONE TAKES place in Bangladesh. Another deep in the Louisiana bayou. Sparks also flare from the eastern shore of Staten Island, after the storm that took Leonard Montalto's life.

Before I moved to the Ocean State, I lived in New York City. Before Miami and Phippsburg there was Oakwood Beach. I was working at the College of Staten Island in 2012, during the fall that Sandy spun into the harbor. Both the size of the storm and its unusual route were unprecedented in scientific memory. Never before had the water reached so high. Of the city's nearly eight million residents, over four hundred thousand were inundated, many of whom lived atop land that had formerly been zoned as tidal marsh. While flooding in these neighborhoods was common, Sandy exceeded all previous experience. In Oakwood Beach the storm surge topped out at a record-breaking fourteen feet. The college campus remained closed for weeks. When classes finally resumed, some of my students were missing, displaced or worse by a previously unimaginable amount of salt water.

One, a brilliant Russian woman named Lena, had been living in a basement apartment in Midland Beach. During the storm the ocean poured into her rented room. The little she had was ruined. Her bed, her books, even her computer; all became bloated with water. I offered her my couch but she said she would stay with a friend. As the semester progressed, Lena stopped coming to class regularly. I don't know if it was the commute from her temporary housing in Jersey or her lack of funds that finally did her in. Either way, she disappeared. A few months later she wrote me a short e-mail from her landlocked home in central Russia, saying thank you and goodbye.

I suppose you could say it was then that I knew that the coverage of the storm and of all that it gestured toward was incomplete. Where was Lena's story? And though I had yet to meet her, where was Nicole's? Where were the stories of those who had been flooded before Sandy? And of those who, in the wake of a storm so powerful it sucked the light right out of the tip of Manhattan, had left?

For much of the last half century, the eastern side of Staten Island was the kind of place where teachers, firefighters, cops, and sanitation workers could have their own version of the good life, digging for mollusks in the mudflats, fishing for stripers off the pier. In places like Oakwood Beach, there were clambakes in the summer, and the neighborhood kids played soccer together at night under the streetlights. Sure there was a flooding problem and a wastewater treatment plant, but it was considered home and a good one at that. Leonard Montalto grew up there and he liked it so much he stayed put, raising three daughters in the little white cottage on Fox Beach Avenue. His sister, Patti Snyder, raised her family just down the block. And when Patti's daughter moved out, it was to a bungalow right across the street from Leonard and his children.

Despite their love for the place that had long defined them, after Sandy, residents of nine local communities began begging the state government to bulldoze their homes and allow the land to return to tidal marsh. This, more than anything else about Sandy and its aftermath, surprised me. Not the fact that Goldman Sachs was one of the few buildings below Chambers Street to keep its power intact through the storm. Not the fires that raged out at Breezy Point or the elderly stranded in the Red Hook Houses for weeks. It was the clamor rising from the sodden side of the city's only Republican borough, the signs that read, "Mother Nature wants her land back" and "Buyout Wanted, Buyout Needed." What did these residents of right-leaning, climate change–denying, low-lying, working-class neighborhoods know that the rest of us did not? How was it that they were interested in retreat, one of the most progressive and controversial adaptation strategies for sea level rise?

When I finally make it out to Oakwood Beach that summer, over a hundred million dollars have been allocated to purchase and demolish the tight-knit seaside community. The work of unsettling the shore has begun.

✳

The trip from Manhattan takes a little over an hour. From the ferry deck I watch the century-old skyscrapers recede. Once on Staten Island, I ride my bicycle down Bay Street through Little Sri Lanka, among the two-hundred-year-old stone cannon mounts at Fort Wadsworth, and out along the boardwalk on South Beach. The bustle of the city starts to fall away. The bike path is suddenly studded with dunes and cedars and black needlerush. An abandoned airplane hangar, a washed-out teal jungle gym, and a stone-gray wastewater treatment plant. I feel as if I am in some neglected corner of the Hamptons, yet I have not officially left the city.

Twenty-two thousand years ago the massive Laurentide Ice Sheet began to withdraw. It had covered New England and all of New York City in nearly mile-thick glaciers. When the ice pulled back, much of the land that lay just beyond its farthest edge subsided, creating hundreds of miles of swamps, bogs, and tidal marshes, including those that line Staten Island's eastern shore. At the turn of the last century, there were over three hundred square miles of wetlands within a twenty-five-mile radius of New York's city hall. Where the land met the sea, muskrats made mischief, white water lilies bloomed, and egrets nested. Neither wholly water nor wholly terra firma, wetlands, at least in postcontact North America, were rarely explored or developed. That is, until the Swamp Land Act of 1850, which gave states ownership over any marsh they could drain. Ever since, these unique ecosystems have been under threat. Land that once was deplored, in part because of the difficulties speculators faced in placing hard boundaries around blurry edges, suddenly provided a chance to make money from something that had been, for the longest time, considered worthless.

As the population of the New York metropolitan area expanded, roughly 90 percent of the city's wetlands were backfilled and hardscaped. Chinatown was once a wetland. Coney Island was once a wetland. East Harlem was once a wetland. So were Red Hook and the Rockaways. Broad Channel, Bergen Beach, and Canarsie. John F. Kennedy International Airport is sited atop former tidal marsh. So are Fresh Kills Landfill and the Brooklyn Navy Yard, a healthy chunk of coastal Queens, and almost all of Staten Island's eastern shore.

It's not just Gotham where wetlands once reigned. Much of the Northeast Corridor, the most densely populated portion of the country, was covered in cordgrass not that long ago. Since the eighteenth century, Rhode Island, Connecticut, New York, and Maryland have all lost over 50 percent of their coastal wetlands to development. Big chunks of Boston, Providence, New Haven, Philadelphia, Baltimore, and Washington, DC, were all once so wet that no one dreamed of living there. These seemingly mundane landscapes were not fawned over or earmarked for preservation. Instead, in urban areas, they often became informal garbage dumps—damp, unprofitable land fit for hiding trash.

Around the turn of the last century, a significant portion of these wetlands turned landfills got paved over to meet the demands of the region's growing industrial ports. Then, as the shipping industry waned in the forties, the mixed industrial areas were redeveloped once again. At the time, living alongside our country's polluted waterways was considered a nuisance, so public and low-income housing often went in. The population boom of the fifties led to a shortage of residential units, and the once-soggy edges of many cities provided cheap, if flood-prone, shelter to those who did not have enough money to live anywhere else. As the century progressed these were also the neighborhoods that didn't receive much infrastructural support; they were the places that flooded most regularly and got the least help.

*

A few months after my first visit to Oakwood Beach, I stop by Alan Benimoff's office at the College of Staten Island. Our resident geologist, Alan has been working on a series of papers in which he attempts to expose some of the underlying causes of Sandy's devastating impact. When I first see him, he is hunched over his computer at the far end of a dimly lit room littered with different earthly artifacts—rock samples, embossed topographic maps, and replicas of prehistoric fossils. Alan lets out a sigh big enough to travel. Then he looks up and gestures for me to come closer. It is an unseasonably warm late-winter day, and the sky beyond his window threatens thunder. The campus should be covered in snow but instead is pocked by mud and puddles.

Potbellied, balding, an old-school Italian American with a big white mustache, Alan strikes me as an unlikely climate change specialist at first. While he is reluctant to talk about the future, he has no problem discussing how poorly planned urban environments contributed mightily to the chaos Sandy wrought. On his computer he pulls up a layered map of Staten Island's eastern shore compiled from various data sets: population density, topographical features, building types, zoning codes. Most of the land is bright red, meaning that it lies no more than ten feet above sea level. Some is shaded light blue, making it difficult to distinguish from the bay. "Blue means the area is zoned as a wetland," he explains.

Alan's map also shows building footprints. He clicks, and the information displayed on the screen changes. "This is the turn of the century," he says. "You can see that the area was mostly marsh, with a few buildings indicated in black." I am surprised to discover that back then the borough had a different shape. It was not the triangle I tend to think of it as being but rather more of an hourglass, with most of the desirable neighborhoods buffered by a belt of arterial wetlands cinched around the island's waist.

Alan shows me the last hundred years of Staten Island's development in ten-year intervals. As the century progresses, the number of black building footprints increases, even in the areas that previously weren't considered land. There the jagged lines that indicate marsh grasses are plastered over, and a street grid emerges. "Wetlands act as giant sponges, absorbing storm surges. When they are paved over, that water still has to go somewhere, crashing into everything in its path," Alan says. "No one talks about it, but the way we have developed the coast amplified Sandy's destructive force."

He looks at me through rimless round glasses and adds one final data set to the map. Twenty-four red dots appear scattered along Staten Island's coast. "I've plotted every single Sandy-related death as well. The important thing to realize is this: over half of the people who died in the storm were standing atop land that once was a tidal marsh. If you ask me," he says, his cursor hovering over the fragile fingers of development that compose the easternmost reaches of Oakwood Beach, "none of those homes should have been built in the first place."

*

After forty minutes of riding I eventually arrive at the edge of Oakwood. I have seen a single building razed before, but nothing prepares me for watching an entire community get wiped off the map. The crunch and snap of backhoes eating away at siding sounds at the far end of Kissam Avenue. One yellow machine mounts a pile of debris and gnaws like a praying mantis dismantling its prey. The farther I ride down the street, the less I hear, because the demolitions are mostly complete, some of the houses already gone.

I lock my bike to a tree so I can move more slowly. Waves of invasive grasses keel around the dozen or so concrete foundations that remain. I walk down what was once a driveway, out to a slab that was once a house. Most of what made this place home in the strictest sense—the walls, the roof shingles, the joists—has been broken apart and now waits to be carted away.

Wind blows in warm scraps while I investigate the smashed-up concrete, the abandoned gutters, and the sheets of Pepto-Bismol-pink fiberglass.

A family of geese waddle across the rubble, then veer off, disappearing into the marsh like soap bubbles popping: one-two-three. I follow them, venturing a little farther into the rambunctious green. The cordgrass and cattails get caught by the wind and sway. I step carefully, feeling out the uneven ground. Red, tannic water wells up around my feet while a zebra finch sings from the broken branches of a nearby tree. It is not my first time visiting a marsh, but it is, in truth, one of the first times that I am really paying attention. The calm that washes over me is immediate, the city's stresses sloughing off in thick sheets. I had expected this day in Oakwood to feel like an excursion to a ruin, but the neighborhood and the surrounding tidal marsh are alive in ways I hadn't anticipated. This place is both accursed and holy, the land forsaken by humans and also in the process of being reclaimed by forces beyond our control. Within this tension, I feel strangely at peace.

For most of my life I never gave tidal marshes much thought, but now they are, in their sly and unassuming way, absorbing my attention. To most, a wetland is just a mess of grass. The sulfuric scent of decomposition. Miasmas and mud. But I am beginning to see them as divining rods, signaling where there will be more water in the future. And even more importantly, that the future is, in many cases, already here.

✳

A year before my first visit to Oakwood Beach and two weeks after Hurricane Sandy, Joseph Tirone, a boyish fifty-year-old real estate broker and longtime resident of Staten Island, made a trip to the FEMA outpost at nearby Mount Manresa. He went in seeking a small-business loan to help him rebuild but left with an entirely different plan: he would attempt to get everyone in Oakwood, where he owned a rental property, to agree to leave. If he could show near-unanimous interest in retreat, the community would have a shot at securing funds for relocation through the Hazard Mitigation Grant Program (HMGP). The federal government would pay prestorm prices for the homes, then knock them all down so the land might act as a buffer in the next storm.

Joe tells me all of this as we drive together down Guyon Avenue, one of the streets Nicole Montalto used to flee the flood-waters. It is a couple of months after my first visit to Oakwood and immediately after my talk with Alan Benimoff. Even though we are passing over relatively high land, many of the homes we drive by are marked. Flood lines ring the lower fourth of these structures like the trace cold coffee leaves inside a porcelain cup. Piles of bloated furniture and family photographs line the side-walks in front of others.

The word *retreat* usually implies defeat, often in a military set-ting. On the Staten Island where Joe grew up, it would not have been a popular disaster recovery strategy. But that Staten Island no longer exists. A waterlogged doppelgänger has taken its place. Over the past decade, as the number of severe flooding events has increased, residents of the "Forgotten Borough" have regu-larly found themselves at the bottom of the city's to-do list. With Sandy, local patience reached its limit. Retreat began to sound like relief.

Joe first brought up the possibility of relocation about a month after the storm, during a meeting on which kinds of aid were available where and for whom. Some folks were living in homes without heat, with mold working its way up the walls and rotting the floorboards. Others were hunkered down in nearby hotels. He asked residents, "If offered a fair price for your home, who among you would be interested in selling and leaving?" Nearly every one of the two hundred people in attendance raised a hand.

"There was a palpable buzz in the crowd when those hands went up that night in the auditorium," Joe remembers, "because no one expected their neighbors to want to leave." Joe would become one of the leaders of the grassroots buyout movement that swept the island like wildfire. Buyout committees went door-to-door, gauging interest, raising awareness, and mapping the areas where residents no longer felt safe living. Despite the fatigue that came with having to learn the countless acronyms associated with disaster recovery, organizers remained energized by the possibility of retreat. "Everyone had a job to do," Patti Snyder, Leonard Montalto's sister, tells me over a glass of sweet iced tea months after my first visit to Oakwood. "I've lived in the community since I was a girl, so I was in charge of getting people on board and tracking down those who hadn't returned after the storm."

While Patti made calls, Joe canvassed the neighborhood, educating those who remained on the rules that governed the buyout process. "We tried really hard not to intimidate people," he says. "This was information that we had to carry into the community. We couldn't make people feel cornered, like they were being forced out." While the strategy was one of group flight, the decision had to be arrived at by the individuals who had long considered Oakwood home.

Joe rattles off the names of the former owners of almost every single house we pass. "That's Danielle Mancuso's old place, and Pedro Correa's, the Iraq vet. Here's where Joe Monte used to live. He built out his bungalow by hand over eleven years, slowly turning it into his dream house. Just before Sandy he put in the finishing touch, a wrought iron fence. His place really was a labor of love. For a long time, he struggled with the concept of giving it all up. But eventually he came around. Apparently they decided not to bulldoze his apple tree, which made him really happy." As we drive past, I see the tree's bare limbs aloft in the wind. But there is nothing else on the lot. No house. No fence. Nothing to call one's own.

Suddenly the sky rips open. Rain starts falling in thick out-of-season sheets. Joe stops the car in front of an 18-wheeler with a billowing American flag painted across its flank. Every Thursday the trailer opens, offering free boxes of cereal and canned tuna to the people who still live out here, the ones who are either hold-outs or waiting to close on the offer from the state.

"One of the biggest concerns was that the land was going to be redeveloped," Joe says. The wipers momentarily make clear the scene the rain blurs. "It was a lower-middle-class neighborhood, and everybody was pretty much at the same level of wealth, or lack of wealth. If their homes were going to be given to a rich person, or if they were going to be knocked down so some developer could build a mansion or a luxury condo, they were not leaving. They'd stay there, they'd rot there, they'd drown there, but they were not leaving."

Fortunately an HMGP buyout mandates that the land be returned to open space in perpetuity, and it was this more than anything else that convinced holdouts to participate. "I can't tell you how many people said to me, 'OK, you're sure this is being returned to nature, then I'll do it,'" recalls Joe. Residents were willing to give up their waterfront private property—a luxury that very few could afford elsewhere in New York City—if, and only if, that property was to become a commons of sorts, the right to use the land held by all.

Rather than viewing the buyouts as a further tool of systemic marginalization, many Staten Islanders began to see them as a chance to finally move away from the long-term neglect that had contributed to their vulnerability, offering an opportunity for a fresh start. "We have been trying to get the city's attention for years. Whenever it rains, the street out front fills with water, and that's just the beginning of it," recalls John Hojnacki, at the Veterans of Foreign Wars clubhouse right across the street from the rescue trailer. After only ten minutes of rain, a gigantic puddle has already formed on the pavement. "It took the city over a decade to install the berm they promised would shield Oakwood. Then Sandy ripped its top right off and yet another surge of storm water rushed in." He flattens the creases in his red bowling shirt and moves toward the door marked "Canteen." The bricks on the wall behind him are two-toned. The darker ones sat submerged beneath the salt water, while the lighter ones didn't get wet. The line separating the two is well over his head.

When I ask about the correlation between climate change and the flooding, John's neighbor, who has walked over to join us, demurs. "I don't know for sure what is causing it, but we've been flooding worse and worse, year after year. If they want to offer me a fair price for my house, I will take it and I will leave." As I look out over the adjacent tidal marsh, I realize that as rising tides peel away financial security, pragmatism trumps politics.

By January 2013, the Oakwood Beach buyout committee had the overwhelming majority of residents interested in retreat. They leapfrogged their local representatives, who they thought would oppose something that could hurt surrounding housing values and much-needed tax revenue, and brought their case directly to Andrew Cuomo, the state's governor. He praised them for coming together to make a difficult decision and later that month announced a pilot program to purchase Oakwood Beach homes at prestorm prices. The program would be voluntary, but anyone in the buyout zone who was interested could opt in. Typically HMGP grants are used to relocate rural riverine communities out of the floodplain, making the funds' allocation in Staten Island somewhat novel. It marked one of the first times in the history of the program that coastal properties in a densely populated metropolitan area would be purchased and demolished.

Within a year of the storm, some residents of Oakwood Beach were closing on the sale of their homes to New York State. Many moved nearby, taking advantage of a 5 percent sale bonus offered to encourage relocation within city limits. Meanwhile, the city's glacially slow, poorly managed "Build It Back" program had yet to cut a single check.

Seeing their neighbors receive much-needed aid, residents of nearby Graham Beach, Ocean Breeze, Midland Beach, South Beach, Crescent Beach, New Dorp, Tottenville, and Great Kills organized buyout committees of their own. "We had been trying to work with local politicians for four or five months and it was getting nowhere," remembers Frank Moszczynski, who lost his Ocean Breeze home during the storm. On weekends Moszczynski and other local leaders took turns tending to a pop-up tent where they distributed information about the retreat. Soon thereafter signs went up urging Cuomo to "Buy the Bowl," a nickname the neighborhood had earned from its tendency to fill with floodwater for weeks. Eventually the governor acquiesced, expanding the initial buyout program twice to include well over five hundred homes.

✳

During my first visit to Oakwood Beach I discover that each of the demolition sites is roped off by bright-orange plastic fencing. When a house has been completely dismantled and every last trace of the former occupant removed, the crew taps a handmade wooden sign into the sandy soil out front, marking the former street number. Down on Fox Beach Avenue, a piece of plywood with the number 89 stenciled on it leans to the left. Next to it is number 87, the former site of Joe Tirone's old rental property. Nothing remains of the tiny two-bedroom beach bungalow.

The blue of the sky is so bright and so clean and contrasts so completely with the scene that it strikes me as ominous. At the near end of Fox Beach Avenue many homes still stand, their windows boarded over. Someone has spray-painted the words "Praise Jesus" over one such plywood sheet. In front of a green midcentury ranch, a small American flag is tacked to a tree. A little down the way, a plank swing sways in the damp air. But no one is there to ride it.

I keep walking and come upon a planned neighborhood at the far reaches of Oakwood, where all the row houses are eerily similar: fake brick facades, cheap plastic garage doors, and matching plastic mailboxes out front. Both housing projects that I have visited today were built in the last ten years, directly on top of an area zoned as a tidal wetland. The developers received variances from the city's Board of Standards and Appeals, exempting them from a 1973 law that forbids the backfilling of coastal marshes. Hundreds of thousands of dollars went into these homes' construction, and now, less than a decade after completion, each of them has been slated for demolition, the risk of future flooding considered too great.

Out back I discover a line of eight identical aboveground swimming pools. Algae bloom green and brown, and minnows dart in and out of the shadows. Each pool is separated from the next by a flimsy white plastic picket fence, buckled where a surge of storm water hit. Like so many neighborhoods across America, this one is a poorly constructed, shoddy version of the American Dream, built for profit instead of longevity.

I knock at the one town house without a demolition notice tacked on the door. A woman answers.

"We're still waiting to be bought," she says. "We submitted the papers but we're still waiting." She is from Russia. Her toenails are painted pink. Her fingernails are painted pink. Across the front of her T-shirt the word *PINK* is printed in capital letters. "You ask if I'm happy? I'm not happy," she tells me. "This used to be a good area. All quiet except in the morning with the birds singing. Now the houses are empty. Now there are raccoons and rats."

"I'm sorry," I say. "That must be hard." But how can I know, really, how hard it is?

✳

Staten Island's eastern shore lies between one and nine feet above sea level. Due to the relative evenness of this part of the borough's topography, a small increase in the overall amount of water in the area can lead to wide-ranging moderate flooding. This is why many of its coastal neighborhoods have both formal and informal storm protection in place. Oakwood Beach is separated from the sea by a ten-foot-high berm. A little farther south, Father Capodanno Boulevard, a slightly elevated shore highway, has long kept the Atlantic out of Ocean Breeze. But a remarkably high tide, coupled with Sandy's trajectory and sea level rise, meant that an unprecedented amount of water was funneled into the harbor's open mouth during the storm. And while berms and raised infrastructure can and do buffer neighborhoods from regular flooding, when breached they tend to exacerbate the problem.

Sandy was a slow-moving storm; water levels rose gradually. Oakwood Beach and Ocean Breeze both sit in topographical bowls, hemmed in by higher ground. So residents had no idea that the storm surges were getting higher, because they were protected. Up until about 7:20 p.m., Oakwood Beach stayed relatively dry. Once the waves were higher than the berm, however, the floodwaters poured in, filling the streets in minutes.

Nicole said, "If I had to guess the time, I probably left around 7:30, and that was when the water was suddenly getting pretty deep."

The last thing her father said was, "The water is rushing in. I gotta go, I gotta go."

The flooding wasn't gradual. It was sudden and violent, like the way the earth's climate changes: jerking back and forth between different equilibriums. Hothouse, then glacial. Dry, then inundated. The transition from one to the other dramatic.

If you are struggling to imagine what happened in Oakwood Beach on the evening of October 30, 2012, take a mixing bowl and place it in the bottom of your sink. Hold the bowl down while the sink fills with water. The inside of the bowl stays dry—that is, until the water reaches the lip, when it comes rushing in. That abrupt transition from mostly dry to anything but caught many off guard. In Oakwood Beach and Ocean Breeze alone, fourteen people—a third of the total for New York City—died during the storm.

"I remember looking out the window and seeing a wall of water coming down the street," says Loisann Kelly, who lived a couple hundred yards from the Montaltos. "It moved down the road, eating everything in its wake—cars, trees, porches."

Loisann's living room flooded in minutes. Her house sits on nine-foot-high stilts, a design precaution she and her late husband implemented when they purchased the place in the eighties. Still the floodwaters rose higher. "I've seen scary things—I saw the plane hit the Twin Towers—but this was worse, this scared the crap out of me. It was the longest night of my life." Loisann floated for hours on her couch in her wine-dark living room, which swiftly and inexplicably had become part of the Atlantic

Ocean. "I heard my neighbors screaming, yet there was nothing I could do but wait." Wait for the dawn. Wait for the water to go down. "I had no idea how long it would be. I think I fell asleep at some point, and when I woke, the water was gone."

The surge would destroy over half the homes on Kissam Avenue. Flattening some and sweeping others off their foundations, dragging them through the cordgrass and depositing their shattered forms in the surrounding saltwater marsh. Patti Snyder's husband would spend hours on his rooftop awaiting rescue. When Pedro Correa's house collapsed, he jumped from it onto the roof of another house that serendipitously floated past. Just down the road, Eddie Perez was able to save himself from the floodwaters by scaling an oak tree. "That night I learned how to hold on," he tells me when I speak with him on his crumbling front steps.

✳

On my way back to my bicycle I pass a white bungalow on Fox Beach Avenue with caution tape draped between two decorative bushes. The bottom floor of the home is buckled, the structure listing to the left. Though I do not know it yet, this is the home where Patti Snyder and Leonard Montalto grew up, where Leonard later raised his daughters, and where Nicole returned, calling out his name.

I pass empty lot after empty lot and stop at each of the three homes with cars still parked in their driveways and lights still on.

Franca Costa tells me that there are disaster tourists who hang out the windows of their rental cars and take photos. Her cottage is decorated with three rainbow-colored pinwheels, six ceramic angels, a rocking chair, three tomato plants, two wreaths covered in plastic flowers, and a handmade sign that reads, "Please do not park in front of the driveway." She is the only person on her street who did not accept the state's offer to leave. Franca tells me that everything is changing. That last year a dead dolphin washed up on the shore, and that her dogs got a bacterial infection from swimming on the other side of the berm and their poop turned to water. That there is seaweed now in August and sand flies and blackflies where there used to be none.

Then she says, "It's like a little piece of heaven. God bless the people who left, but I can't. I just can't start over like that. I don't have the money. I still owe too much on this house, and regardless of what they give me I won't be able to be so close to the water anywhere else." She hands me a PowerBar and wishes me well on my journey. "Come back anytime."

I wonder if, once most everyone and everything is gone, the city will stop repairing the street after it floods. If the S76 bus will quit running. If the electricity will be restored after a storm. How long will Franca's house be worth the hundreds of thousands she paid for it back in 2002? And after the value leaks out of every last wet lot, how long before the marsh grass reigns? And how long will that reign last? Eventually the ocean will lap higher and higher up the stalks, until they too disappear beneath the water's surface.

I turn the corner and come across a man in a white T-shirt and jeans carrying boxes out to an 18-wheeler. If he stays, he thinks, soon the house won't be worth anything. "I have no choice. I give the city my home on Tuesday," he says, turning back to the task of transferring his life into a long metal box.

On the corner of Fox Beach Avenue and Mill Lane there sits the empty socket of a demolished home, the foundation filled with water and floating bits of Styrofoam.

"I feel that it's time," a woman on Promenade Avenue tells me. "I'm moving away. Every night my husband would ask, 'Should we do this?' The last couple of months were sad. There used to be fifteen kids at the bus stop and now there are only two. They used to all wave their hands and now they don't."

✳

Here is what I found among the rubble-strewn foundations and the boarded-up bungalows, alongside the berm with its insides exposed and the inundated cordgrass. You might say it was yet another moment, in what was becoming a long string of moments, when I started to look at the ground instead of the horizon for a glimpse of what was to come. It is no surprise that there I found another tremendously vulnerable tidal wetland. But in Oakwood I began to understand that the vulnerability of these places can and ought to be transformed into a battle cry. Yes, wetlands communities are the canaries in our coastal coalmines, the first to feel the ocean's gathering force. And yet the retreating residents of Oakwood, by banding together and demanding aid, are also something else. An example for the rest of us to follow. Lights along the landing strip, illuminating the way. They are less victims than agents. More rhizomes than rampikes.

Cordgrass takes its name in part from its vast rhizomatic root system. The ancients knew that both the creeping rootstock and the blades of this plant could be woven into a strong rope. A knotted rope, to measure the distance between places. A rope painted red to mark and round up citizens for public meetings. A twisted rope to launch an early catapult. This is the derivation of its genus name, *Spartina*, "cord" in ancient Greek. Like its homonym, *chord*—three notes played together in the braid of sound that makes a harmony—a cord derives its strength from its weave. Speak either this plant's common name or its scientific name and its potent anatomy manifests in the air.

Cordgrass's subterranean network of rhizomes is why it is difficult to dig in a healthy marsh. The web of connective tissue running through the soil is dense and strong. But when cordgrass is impounded with stagnant salt water, the plant's rhizomes retract. The sediment around them loosens and the ground literally starts to collapse. But the transformation doesn't stop there. What distinguishes rhizomes from normal roots is that they are not simply reactionary; they do not grow only downward from the plant base seeking nutrients. Rhizomes, it can be said, have a mind of their own. They find the line of flight and act. When the plant is threatened by too much salt, for example, horizontal root growth often starts reaching steadily uphill, away from the element that will not suit. If there is space for the marsh to migrate, it will. From each root a new shoot sprouts—the community, and the home it provides, remade from within.

I pause when I reach my bicycle, leaning against the trunk of a rampike. I gaze past its bare, prophetic form, out into the marsh. Even though I cannot see it, I know that the arterial network of cordgrass rhizomes beneath the surface is both retreating and reforming. Attempting to find not only a way forward but also a way to continue to be itself, albeit in a slightly different location. Just beyond the arching marsh grass fronds, a line of homes waits to be demolished. A moving truck abuts a bleached ranch with a stained-glass door. Its hollow body filling with sofas and family photos, pots and pans and soccer balls. The rhizomes, I think, are not the only ones who know the time has come to withdraw.

Just then a bone-white egret flies overhead. It beats its wings many times, then banks hard to the left and coasts out over what I no longer know how to name. Out at the edge of the marsh a white plastic bag is caught in the reeds. It flaps like the flag of willful surrender.

On Vulnerability

Marilynn Wiggins: Pensacola, Florida

WE'RE FLOODING WORSE NOW, BUT GIRL, THE FLOODING has always been bad. And because we're a black neighborhood the city doesn't pay us no mind. Any house that sits on the ground, that's on slab, those people flood out all the time. We're surrounded by water. The Gulf of Mexico is right there and we've got the rainwater coming at us from the other side. A long time ago, I hear this was a wetland. Back when I was younger I used to drive school buses for a living and there were many mornings that I couldn't get to my bus because the floodwater was so high. I would walk out my front door and the water would hit me at my waist.

Plus on top of it all we had the old sewage treatment plant down here and the mosquito control plant. They were handing out poison. I mean they made high-powered chemicals right around the corner. There were toxins in the soil and underground at

Corrine Jones Park, right here on Intendencia Street. So the people who owned the mosquito plant had to pay $250,000 to the park. Some of the nearby residents were complaining, saying, "How's it possible for the poisonous chemicals to travel from the old plant across the street to the park without our homes being affected?" But we haven't been able to get Ashton Hayward, the mayor of Pensacola, to come see what is actually going on out here. My neighbors next door complain that they have black dust coming out of their faucet. But we haven't heard from the people who test the water yet. It seems like I'm always waiting for someone to get back to me.

Back in 2014, the flooding was very serious. My house is six or seven feet higher than my neighbors', and even so the porch was ripped off. But next door, I tell you, the ones with the half-dead cypress out back, they were flooded *out*. The water went into all those cottages and halfway houses and those people lost everything they had. Many of them just left. The worst part is that when there's a great flood you can still smell the sewage—you would think that the plant was still up and running. But it's not. It closed after Hurricane Ivan.

Across Pensacola, in a lot of black and minority neighborhoods, they're placing retention ponds to try to control, at least a little, the water that's here. But you have to remember that these retention ponds are also a concern for a lot of residents. When they replace the parks where the kids used to play, it can be devastating. Of course the retention pond is needed but it was needed forty years ago. Where were they forty years ago? Can someone answer me that?

I'm sixty-one years old. I didn't grow up in the Tanyard. When we moved down to this corner of Pensacola I had to be in the seventh grade. This was a bad place to grow up as a black woman because it was a red-light district. There was a motel right on the corner called Pope's Place, where everyone went. I remember getting run home many days by white men mistaking me for a prostitute. My father, he even dressed up in dresses just to scare them away. I can still see him standing on the porch, in my mom's clothes, hollering.

Now I'm the president of the Tanyard Neighborhood Association. I hear all about what goes on in this area. Those neighbors that flood out, if they're in government housing, then they have insurance through the city. But it's a hard life situation to have to redo your home every time it floods, and that's what you have to do if you have one of those policies. You have to come back to the place that flooded and start all over again right there. The new neighbor in the new house across the street, she has to have flood insurance because she is paying a mortgage. If something goes wrong with that house the bank wants to be able to get their money back. For me, I don't have a mortgage so I don't have to have flood insurance. But I hear that there's a possibility that we all are going to be forced to purchase it at some point in the future, and I don't know what I would do then because I don't think I can afford it.

Risk

Pensacola, Florida

THE WARM, DAMP RAG OF EVENING HAS SETTLED OVER
Pensacola Bay by the time I reach Alvin Turner's double-wide.
His trailer sits about a quarter of a mile from the water, in one
of the city's lowest-lying neighborhoods, the Tanyard. Long
before the Tanyard was a mostly black residential community,
it was the place where animal pelts were treated before being shipped to Europe, and before it was a tanyard it was a
tidal marsh. The Tanyard no longer *looks* like a wetland, but it
hasn't stopped acting like one. When there is rain, the runoff
from the surrounding neighborhoods all flows downhill, collecting here. When Hurricane Ivan spun into Pensacola Bay it
brought 130-mile-per-hour winds and a storm surge so powerful the sewage treatment plant was destroyed. No matter
what kind of weather arrives—tropical depression or thun-
derstorm, cyclone or hurricane—you can bet that the Tanyard
will be underwater.

I pause on the wooden steps that connect Alvin's trailer to the ground. A metallic ache is growing in my mouth where my last molars sat until a week ago. My tongue still thickened from the surgery, I swallow two aspirin and knock. When Alvin answers, eerily cold air flows out of the cavernous room behind him. His eyes are rheumy—the mucus color striking against the dark interior and his dark skin—and he is wearing a pair of blue plaid pajama pants with a long gash up the right leg. On his shin an infected wound festers. I am ashamed to say that for a second I consider turning around—that for a second I am afraid of this poor, elderly, ailing, black man. As a woman and a nonfiction writer whose work regularly involves interviewing strangers, I have developed one rule: read the situation, and if something doesn't feel right, leave.

But as a white woman and a nonfiction writer, I also know that I have blind spots, biases, responsibilities—and that these things can interfere with each other. I know that simply walking away is a privilege not always available to my subjects, those who make their lives along our shifting shore. And I know, even in a moment of fear, that what might feel convincingly like instinct is actually learned and poisonous. So I extend my hand, asking to be met halfway, and we shake.

Soon I am navigating through the gloom of his living room, where the evening news is the only source of light. Images of a flood in nearby Baton Rouge flow across the television screen: water up to windshields, families in skiffs searching for those who did not get out in time, aerial shots of the mud-brown Mississippi River and the live oaks and the cookie-cutter roofs all mixed up in one another.

"Thousands left homeless after devastating floods in the South," the announcer coolly intones. A radar map shows the bright red of this nameless storm, the frayed edges pulsing in and out like the perimeter of a wound, the scab forming and retreating. Alvin settles into the couch and gestures to a love seat nearby. I pull out my pad and pen.

"Ivan tore up so much of my house that me and my wife had the old one knocked down. And now I got this here," Alvin says.

By now I know that many of the absolute lowest-lying areas—the places that flooded most regularly even when the ocean was more static—have historically offered shelter to those who literally couldn't afford to live anywhere else. Wetlands have long been viewed as wastelands. The Hippocratic writings that incorrectly link the stagnancy of the marshes' waters to the production of phlegm echo in the cries of "Drain the swamp!" ringing through the streets of Washington today.

In the antebellum South, wetlands were thought to corrupt the air and whoever breathed it. William Byrd, an early land surveyor in Virginia, describes what would later become known as the Great Dismal Swamp this way: "The exhalations that continually rise from this vast body of mire and nastiness infect the air for many miles round. . . . It makes [nearby inhabitants] liable to agues, pleurisies, and many other distempers that kill [an] abundance of people and make the rest look no better than ghosts." Which is why runaway slaves and displaced Native Americans sought out the Swamp and other wetlands as refuges, constructing many of our country's Maroon and Indigenous communities in the marshes and bogs that line the East and Gulf Coasts. They threw down roots along the damp fringes of this country precisely because these places were easy to defend and difficult to attack, the land itself less than coveted.

And so what I once thought of as inquiry into vulnerable landscapes—and the plants and animals that call those places home—has also become an inquiry into vulnerable human communities. I've come to Pensacola to learn more about an arm of

FEMA known as the National Flood Insurance Program (NFIP) and how it shapes life near the wrack line. The insurance industry relies on bad things happening, but to only a few folks at a time and in a somewhat calculable manner. Flood claims, however, often come in all at once, both unpredictably and in staggering numbers, and as a result private agencies have historically refused to offer this kind of coverage. The NFIP was founded in 1968 to address this gap and defray the expense to the federal government. Residents would finance their future recovery by paying a yearly flood insurance policy premium. Flood maps were drawn up, and if you had a mortgage and lived in a flood zone you were required to purchase insurance.

Since the policies the NFIP sold were both new and mandatory, it offered them at highly subsidized rates, with homeowners paying as little as one-tenth of the actual cost to insure against the risk. In other words, the NFIP inadvertently made living in the floodplain seem much safer and cheaper than it actually is. Over the half century that followed, the number of units in flood-prone areas has increased by a factor of four. Today roughly fifteen million homes sit within "special hazard flood areas." On the East Coast alone, over a trillion dollars in property assets rest alongside our increasingly unpredictable ocean. And every time one of these homes floods, those with insurance are required by law to use their policy payouts to rebuild in place, even when that place has been underwater repeatedly in the past.

After only a few minutes I ask Alvin to tell me about every time a hurricane hit, every time the neighborhood was inundated. I ask him to recall when he had to evacuate, what he lost, how he was able to rebuild, and why he keeps coming back. As I probe, I feel as though I am pressing the gash on his leg and asking how much it hurts. "I'm sorry," I say, and mean it. Sorry for digging into the fertile soil of your pain. Sorry for asking you to tell me how the floodwater chewed through the foundation, how the wind removed the roof, how little there was to come back to. And, in truth—thinking of the moment earlier—sorry for purchasing my plane ticket without contemplating how extraordinarily easy it is for me to visit a place so far from where I am from, knock on a stranger's door, and expect to be let in. Sorry for momentarily fearing you, the man I met when I arrived unannounced, because all I could see at first were the differences between us.

"I'm sorry," I say again. Alvin cringes. There are other ways forward and I am searching for them. "I'm listening," I offer instead, leaning in.

Alvin continues. "I've been living here forty-some—no, fifty—years, since way back. We lived here, my wife and I, because we could afford it. We raised our children here. My granddaughter helps me take care of the place, because my wife, she passed two years ago." An empty feeling seems to fill the interior like a scent, plastic and frostbitten. "I'm retired now," he adds.

The longer I sit on his love seat the more fully my eyes adjust to the lack of light. A Virgin Mary clock comes into focus on the far wall, a scratched Formica countertop hovers in the door frame. *This man will not hurt me*, I think. Alvin is not dangerous, just remote—accustomed to the absence of other human beings, and to living alone on the edge of a neighborhood threatened from all sides.

✳

In my mind, writing and reporting about people—especially vulnerable ones—is an act of empathy. But how to practice that empathy is less clear. In *The Empathy Exams*, Leslie Jamison writes, "Empathy is always perched precariously between gift and invasion." And, "Empathy isn't just remembering to say *that must be really hard*—it's figuring out how to bring difficulty into the light so it can be seen at all." She writes, "When bad things happened to other people, I imagined them happening to me. I didn't know if this was empathy or theft." Perhaps her definition that resonates most fully with working the flood zone is this one: "Empathy means realizing no trauma has discrete edges. Trauma bleeds. Out of wounds and across boundaries."

✻

Two days later I stand with Samuel, a senior colleague, at the edge of the Gulf of Mexico. We drove out across the Bay Bridge, which connects the mainland to Gulf Breeze and Gulf Breeze to the beach. Locals call this flimsy spit of sand the Redneck Riviera, partly because it draws vacationers from Alabama, Mississippi, and Louisiana, and partly because self-deprecation borders on art in the South. The sand is as white as any that I have seen, and it squeaks beneath my feet as I walk from the changing rooms out to the sea. The tiny granules are pure quartz from the Appalachian Mountains. Bits of larger rock that broke apart and were polished during the twenty thousand years they tumbled down streambeds to the edge of the continent.

Samuel pulls at the strings of his swimming trunks, tightening his slouchy board shorts, then steps forward, placing his clammy hands on my shoulders.

"Hold on," he says, turning my body away from him. Then he reads aloud the E. E. Cummings quote I have tattooed on my back, the final line of which reads, "will never wholly kiss you."

Suddenly I feel his damp lips pressing into my skin, into the letters inked there. My stomach slams into the roof of my mouth, locking my words in instead of out. And my body takes over. It wants to punch him but knows it shouldn't. Instead it walks straight out into the ocean, into a flotilla of spawning jellyfish at the northernmost edge of the Gulf. It starts to swim.

✳

Samuel was not the only man to harass me in the Panhandle, nor was this moment even close to the first time someone had made inappropriate advances while I was conducting research. But just because being masturbated in front of, getting my ass pinched, and having my mind filled with lewd comments are regular parts of my work life doesn't mean I wasn't unsettled when one of my Pensacola interviewees told me, "Watch out walking along Palafox Street at night. I might just kidnap you." And when I didn't return his phone call, he left a message on my answering machine, his Bud Light–slurred voice saying, "Where did you go? I have the FBI out looking for you, and when they find you, they're going to strip-search you."

Samuel was with me in Alvin's trailer; at that point, his company felt like a form of protection. He made it that much easier to choose to enter a stranger's home. Before crossing the threshold, I calculated the risk—something that Samuel himself is an expert on. He studies the decision-making processes of at-risk people, businesses, and governments. And by looking at the way that homeowners decide—or not—to carry insurance, in concert with the kinds of coverage the industry offers, he creates policy recommendations aimed at reducing losses—both physical and financial—from the mounting threat of natural disasters.

These days, Samuel's work has become increasingly important. The NFIP was already $24 billion in debt after Katrina and Sandy. Then Hurricane Harvey dumped over one trillion gallons of water on Houston (much of which was erected atop a former swamp), pushing the program past its $30.4 billion borrowing limit. Early reports suggest that more than 450,000 people will file for federal assistance after Harvey, plunging the flood insurance program in even deeper over its head.

If the NFIP is going to survive, something has got to give: the policies will need to become more expensive, more people will have to carry them, severe repetitive loss properties will be bought out instead of rebuilt, or private insurers will choose to get involved thanks to the rise of a speculative financial instrument known as reinsurance, or insurance for the insurers. In every single community I have profiled, rumors about the future of the NFIP circulate with a disorienting force similar to that of the storms from which they were born. On the Isle de Jean Charles and in Oakwood, residents told me they had heard their premiums would rise from two thousand dollars a year to fifteen thousand, should they stay in place. In Rhode Island, people talked about the possibility of flood zone expansion; in Pensacola, residents who own outright fear that the policies will soon become mandatory, even for those who don't carry a mortgage. While versions of each of these options are eminently possible, the future of the NFIP itself remains unknown since Congress will need to reauthorize the whole indebted endeavor in March 2018.

If homeowners do start paying insurance rates that reflect the severity of their risk, the long-term implications will be profound, from property devaluation in coastal neighborhoods and an attendant drop in taxes collected to the potential pricing out of low- and middle-income homeowners, which would further compound economic inequality. Should rates rise, almost every single person I have met along our country's soggy shore would have to leave—not because of the flooding itself, but because of the higher cost of carrying a policy.

This issue is why I decided to learn as much as I could about the NFIP and the different paths the program might take in the future. But the more I learned the more convoluted the whole discussion became, until one day I decided to consult the expert in risk calculation himself, Samuel. His administrative assistant set up a call; I told Samuel about the homeowner interviews that I had been carrying out in wetland communities around the country, and he began to explain the complications swirling around flood insurance reform. When he said he might be heading to Pensacola, I replied, "Me too." We hatched a plan to research the subject together. I hoped my time in Pensacola would help me to turn an idiosyncratic, jargon-filled discussion into one that incited empathy even from those who lived far from the coast. Because, as I was beginning to realize, just as the very shape of our shoreline was transforming, so too were the policies that dictate who gets to live there and on what terms.

Six months later I was sitting next to Samuel on Alvin Turner's love seat.

Samuel. Who suggested we swim in the Gulf before leaving Pensacola. Who asked me to hold his luggage in my hotel room instead of leaving it at the front desk. Who, on our way to the beach, offered me a "senior fellow" position. Who invited me to present at the National Academy of Sciences. His flattery and his grant money. His federal connections and his reputation. Samuel—the expert in calculating risk.

✻

Whenever I tire of reading articles about the National Flood Insurance Program, I dive headfirst into researching Pensacola's past. Soon I discover that before Pensacola was the Redneck Riviera, it was the site of the first year-round European settlement in what would become the United States. When the Spaniards controlled this area they thought it was going to be one of the lodestars of their "New World" empire, an international seaport surrounded by cypress swamps that would provide ample timber for shipbuilding. But Pensacola never developed into the city they imagined. The area's salt marshes, the shallow and short tributaries that reached only so far inland, its difficult-to-navigate bayous and otherwise sandy soil, which was unfit for the production of cotton, meant that Pensacola remained an outpost in the South as opposed to a nexus of activity.

Still, Pensacola has long been a kind of destination spot. Not just for those seeking the South's whitest beaches or the waist-high breakers crumbling out at the Cross. Long before Peg Leg Pete's started slinging rummy Bushwackers, long before the Tin Cow served its first spiked milkshake on Palafox Street, this spit of land was a swamp. And that swamp was, for many, a soggy fortress. A landscape that sided with the dissenters who sought it out.

Desperate to bolster the population and fend off attacks from the other Europeans claiming power, the Spanish encouraged slaves from rival colonies to flee their bondage and head south. They promised freedom in exchange for military service. In this way, La Florida would become a beacon for runaways and home to the first free black settlement in North America. In that place and time when one world was ending and another was being built, Pensacola and its surrounding area were as close as any to that most intractable American myth of the melting pot, with blacks, whites, and Native Americans mixing frequently. The free blacks of Pensacola worked as translators, guides, and trackers; they built and maintained trading posts, captained ships, served as soldiers, and worked in the tanyards.

Pensacola's early fame as one of the most racially mixed cities in the South did not, however, stop it from eventually becoming extremely segregated. By the end of the nineteenth century many of Pensacola's affluent white residents moved to the North Hill and East Hill neighborhoods, located on a ridge overlooking the downtown and removed from the regular flooding. Meanwhile blacks, Creoles, and Native Americans took over Long Hollow, a former streambed, and the Tanyard, where Alvin's trailer now sits.

Vulnerability is inherited, like a garnet necklace or a debt. Here the water comes often and rots what it can touch. When this is your land, you acquire the ruin that comes with it, the knowledge that the sea pulls paint from siding in thick, shiny strips.

※

I lean forward, look Alvin in the eye, and ask if he carries flood insurance.

"Yes," he says. "I'm insured."

"I understand," Samuel offers, "but I'm wondering if you specifically have a policy that covers flooding. It's separate from homeowner's insurance." According to our maps, Alvin's double-wide rests in a high-risk flood zone where insurance is required if you have a federally backed mortgage. Though little is done to enforce the law.

"I don't need any more insurance. I can't afford it," Alvin growls. He thinks that we have deceived him. That we are insurance agents and not the researcher-and-writer team we purported to be. "When you live on a fixed income like me, it takes all the money I have just to keep my bills paid—lights, water, electricity, television, all that stuff. I can't afford nothing else."

"If you don't mind my asking, how much do you make?"

Alvin stutters, then stops. "As I said before, I'm retired. I get Social Security, that's it."

"And how much is that?" I ask.

"I'm not going to buy anything more," he insists.

"You can't afford it, that's what I'm hearing," Samuel says.

"I used to pay $500 a year for my insurance and now I pay $605. I live on $1,300 a month. That's it," he says. The most basic flood insurance policy, added to the homeowner's insurance he currently carries, would cost a little more than one-tenth of his annual income.

Samuel looks at me and raises his eyebrows, as if Alvin's meager earnings are news. I want to ask what he had expected to hear, but I keep my mouth shut.

Meanwhile the news anchor reads from his teleprompter, "The Louisiana floods have crushed the previous records, with thirty inches of rain falling in twenty-four hours, and are causing scientists to wonder if this is just a rare occurrence or if it is likely to become the new normal." There are more shots of more brown water and more people in rescue boats, some wearing life preservers, some not. The water nearly covers a ONE WAY sign on the street. The vast majority of those flooded during this record-breaking nameless storm, I will later learn, are, like Alvin, uninsured.

He looks at the television screen and recognition registers on his face. Alvin leans back. His eyes nearly close. "We got water like that two years ago. I know how hard they have it," he says, changing the subject.

The April 2014 flood that Alvin mentions was caused by unprecedented precipitation, as in Baton Rouge. But the flooding in the Tanyard was worsened by the neighborhood's proximity to the sea. The vast network of concrete tubes running beneath roads and parks, beneath sidewalks and supermarkets, beneath the trailer where Alvin now sits, are designed to funnel rain into the nearest body of water. But when higher tides cover drainage outflows, the inland runoff backs up. Two years ago, when as much as two feet of rain fell in twenty-four hours, the Tanyard turned into a bathtub.

I am told that the amount of rain that fell on Pensacola that day was so uncommon that events like it are statistically supposed to occur only once every five hundred years. Eight hundred and forty *days* later, the intense precipitation that drowned Baton Rouge was dubbed a thousand-year storm. And a year after that, Houston was inundated during a thousand-year hurricane. In a little more than three years, residents of the Gulf Coast have seen millennia's worth of ruinous water.

"I been there, and I made it," Alvin continues, as figures wade through the waist-high floodwaters onscreen. He scratches the wound on his leg and pushes his hand through his thinning black hair. It is the first time in my life I have watched an interviewee as he recognizes his own circumstances in the television's spectral glow, in the life of another whose specific experiences are not his own. I wonder if Alvin thinks of those suffering in Baton Rouge as members of his tribe, as sharing, disproportionately, in the condition of being at risk.

Risk is a word with more than one definition. *Merriam-Webster* says that risk is a situation involving the "possibility of injury . . . [and] peril" and also "the chance that an investment (such as a stock or commodity) will lose value." One definition is physical, the other fiscal, and lately I have been thinking that the difference between the two is a question of proximity. To be "at risk" means occupying the space of the threatened body, drawn close to danger. If peril is primarily financial, however, the person assessing the risk is most likely standing at a safe remove, far from the flood lines. From a distance, risk looks like something that can be managed, through informed decision making or insurance. Where Samuel investigates natural disasters from this perspective, I am more inclined to consider the former definition first.

Over the past half century, our collective perception of the

kinds of risk posed by flooding has undergone a profound transformation. Rebecca Elliott, a professor of sociology at the London
School of Economics, writes that before the advent of the NFIP,
floods were considered "unfortunate event[s] that [could] be
neither foreseen nor prevented. Those afflicted by [floods were]
blameless victims, facing misfortune that might befall anyone,
even those who had made the 'right' choices." We used to think
of someone who flooded out as being exposed, unfairly, to a certain kind of unpredictable and unwieldy weather, as suffering an
"act of God."

However, when the NFIP began mapping flood risk zones
and conducting probabilistic risk assessments, flooding became,
as Elliott puts it, a "scientifically foreseeable, patterned event."
What one can foresee, one can prepare for. And so individuals
were expected to "account for and manage the costs of living in
a floodplain" by purchasing insurance. Put another way, today, if
people are uninsured, they are perceived as having participated in
their own undoing.

Ironically, the more information we have about the likelihood of flooding events the less likely we are to consider those
most "at risk" as being deserving of aid, even when their vulnerability has not been arrived at by chance, but as a result of
centuries of risky and inequitable development. For the first
thirty years of the program, this mattered less than it does
today, as the NFIP was able to subsidize policies and offer
rather robust recovery funding, often through FEMA, even to
those who had gone uninsured. But with eight out of the ten
most expensive hurricanes ever having washed ashore since
2000 and the program teetering on the brink of collapse, our
perception of physical and fiscal risk, security, and who is deserving of that security increasingly determines who gets to
recover and where.

All I can offer Alvin is an exchange. For the sharing of his story, I give him the knowledge that it has been heard by one person. Someone who is very much aware that those at the highest risk of flooding are often also those who can least afford to protect themselves, and that they likely inherited their vulnerability from ancestors desperate to survive.

"With Ivan, we lost everything, the roof was torn off and the water came in and the house was pulled across the lawn over to the other side," he says. "All they gave us afterward was a blue tarp."

"And how did you rebuild?"

"Out of our pockets. And the trailer still isn't fully paid off. That was over ten years ago, I suppose. I'm way down on it, but not done just yet."

When Alvin says this, I know for certain that he doesn't have the additional insurance policies to protect against wind or flood damage—and yet he understands what he risks all too well. His grass lot is permanently soggy. The groundwater is contaminated. And the slow and ongoing work to put in a retention pool nearby, to help alleviate the flooding, was recently halted for fear the displaced soil contained toxins.

I think again about Leslie Jamison, who writes, "Empathy isn't just listening.... Empathy requires knowing you know nothing." So I say, "I can't imagine how difficult it must be for you." By *it*, I mean all of it: winds unknotting the window frames and the rebuilding that followed; the death of his wife and the debt; the noticing for decades that his neighbors often lose everything when they flood, and that some just leave because they don't have the money to rebuild or insure. By *all of it*, I mean not knowing when the next storm will hit or whether he will be able to afford to return. I mean opening the door to my white face, with its passing fear; always being the one who is expected to welcome in outsiders seeking answers. I mean everything I can imagine, and everything I can't.

After five more minutes I thank Alvin for his time, stand up, and move back toward the door and the promise of dinner.

Samuel says thank you too: "You've been extremely helpful. Your participation is *absolutely* vital to our study. We're in a position to make some suggestions to the local and federal governments so that people like yourself can get insured." I want to roll my eyes, thinking that this is likely not the first time some guy in a suit has dropped by for thirty minutes, promising that everything will change thanks to his brief visit.

Outside the trailer, night is gathering and the bats are swooping low. I am standing just where Alvin stood in the last storm. The sea lapping on the sides of a trailer turned tin can boat. Perhaps he looked out at the street covered in dark water and tried to assess the risk that came with evacuating and that which came with staying put, knowing full well that no amount of insurance would have protected his body.

✳

Later that summer one of my students writes me a letter. Zoey is abroad, on the most prestigious research fellowship awarded by our small liberal arts college. While interviewing people in a centuries-old farming community where lives and livelihoods are being transformed by climate change, she runs into a problem. One of the men she gathered testimony from starts sending her texts, declaring his love for her. And when she asks him to stop he gets angry. Like any good researcher she scours the Internet for his name. Zoey turns up a history of sexual assault and methamphetamine addiction.

First and foremost, I respond, be concerned for your safety. Change hotels. Reach out to your support network. Make daily check-in times when you shoot off a text so a particular person knows you are safe. The second bit of advice I offer is this: "The *unwanted* and *uninvited* attention you are receiving from this man who is fifteen years older than you is absolutely not your fault. You are an excellent interviewer and writer and you are doing your job well. That this man is pursuing a relationship with you is not a sign of your sending the wrong signals but of his inability or unwillingness to read them." I want to fly across the Atlantic and castrate her perpetrator for threatening to steal from Zoey the very qualities that make her one of the most gifted young writers I have ever encountered. And yet, I wonder, why am I so ready to defend her openness, empathy, and curiosity when I sometimes wish to scour those same traits from myself?

Jia Tolentino, a staff writer at the *New Yorker*, recently described the aftermath of sexual harassment and assault like this: "One of the cruelest things about these acts is the way that they entangle, and attempt to contaminate, all of the best things about you." What you once thought of as strengths are twisted into weaknesses: if you are open then you are not good at delineating boundaries; if you are empathetic then you are easily manipulated; and if you are curious and friendly then you are asking for it.

When we both return to campus, Zoey and I talk over everything during lunch in the cafeteria. First we talk about her Fulbright application, then we talk about the book *Voices from Chernobyl*, and finally, when half of our time has been used up, the conversation turns to the thing we least want to discuss.

I think about the staggering number of my friends who have been assaulted, either on assignment or by men they knew and trusted. I think about the video that has surfaced in which a reality TV star turned presidential candidate has said that he "can do anything" he wants to women, including "grab[bing] 'em by the pussy." And how it made news for about three days, and then everyone was tweeting about Ken Bone and body language during the debate.

"You didn't do anything wrong," I finally say, rearranging the few previously frozen peas that remain on my cafeteria plate.

"I just wish I'd been better prepared," Zoey says. Since the summer, she has cut her previously shoulder-length hair to within an inch of her skull, and today she wears buckled leather boots and a down vest.

"In what ways?" I ask.

"I wish I'd known that it might happen. And that it would spiral me into"—she pauses, looking for the right word—"self-doubt."

"I should have told you," I say. It happens all the time.

Together Zoey and I start talking about holding a roundtable with other female nonfiction writers and anthropologists to discuss what it means to be a woman carrying out fieldwork. It is a topic, she says, that has not come up once in her four years of studying anthropology. When we start brainstorming about how best to protect yourself in the field, our list of tactics is surprisingly short: Dress like a grandmother. Reiterate the seriousness of your work. Get business cards. Get trained in self-defense. Carry mace.

Zoey realizes the futility of the exercise before I do. "Maybe the emphasis shouldn't be on creating some kind of guide for how to avoid being harassed or assaulted. It seems like an unrealistic goal."

We both had tried to follow our list of instructions and we both had been victims. Our preventative measures had not protected us. There is no insurance we can purchase against the act of rape, against the hand lifting our skirts, against the kiss on the seashore or the threatening text messages; there is no piece of paper to make me forget that a middle-aged white man thought it was appropriate to threaten to strip the clothes off my body and reach inside.

In the moment I first stood in Alvin's doorway, I believed—if briefly—that he was the risk and Samuel a feeble form of protection, and the exact opposite had been true. The more I sat with this knowledge, the more I felt that I had begun to understand the perverse nature of risk: That those considered at risk are taught to fear or distrust each other, instead of those who stand to lose the most should the edifice of white male power crumble. That this distrust often gets in the way of forming meaningful bonds with potential allies. That those at physical risk live in a fiscal risk world where we do not choose the rules or stakes, where our property is always more highly valued than our bodies. That living in my vulnerable body had taught me to fear for my safety, but not how to identify the greatest threat to it. The violence I have been taught to guard against by so many cultural products, from tone-deaf tech advertisements to *King Kong* and *Cops*—where white women are threatened by black men or their avatars—has never, not once, come home to me. But its reverse, where white men regularly, insidiously claim ownership of my white body, has happened so frequently that I've lost count.

In epidemiology, belonging to a "risk group" means being a part of a community with higher-than-average odds of contracting a disease. Historically, however, that label has been misconstrued, singling out that group as posing a risk to others, as being the potential source, rather than site, of infection. For example, the violent stigmatization of homosexual men during the AIDS epidemic. A term intended to signify vulnerability, the need for additional care, instead isolated—and to a certain extent made monstrous—the very people it was supposed to protect.

Alvin's experiences are not synonymous with—are more damaging than—my own, but we both exist in a system where our bodies are often subject and, as such, susceptible to a seemingly endless stream of external aggressions. It doesn't need to be this way. James Baldwin writes: "That victim who is able to articulate the situation of the victim has ceased to be a victim: he, or she, has become a threat." Being "at risk" means being risky—not to each other but to the longevity of an unjust society whose governing principles, social norms, and laws were not, generally speaking, written by those who know, intimately, the fear that comes with physical peril.

They were written by those who know, as Tolentino writes, "that [they] can tie your potential to your female body, and threaten the latter, and [that] you will never be quite as sure of the former again." Written by those who know that whatever their wrongdoing they are less likely to be incarcerated as a result. Written by those who ask to develop in wetlands because they do not have to fund the recovery process; written by those who sign variances and waivers to allow the otherwise illegal construction of (often) low-income housing in the heart of high-risk floodplains. Written by those who earn significantly more for every hour of their time than anyone else. Written by those who, should they happen to be close to the storm's wet center, are able to get out in advance, and for whom the cost of renting a hotel room and eating at restaurants is less of a burden.

Written by those whose power, in its various shapes and forms, keeps their bodies safe. From experiencing harassment, assault, or rape as the price of upward mobility. From living in a community that with each flood is split in half, then split again. From wind; from chemicals blossoming on the water's surface, then settling mutely into the soil; from the storm's warm tide and the darkness that follows.

My first evening in town, before Samuel arrives, I walk through the Tanyard alone. I had been given a list of all the uninsured people in the county who received some federal aid to help them rebuild after the 2014 flood. That aid came with a hitch: once you accepted it you were required to enroll in the National Flood Insurance Program, though the state would cover half the cost of a basic policy for the first three years. Two and a half years have lapsed since the storm when I arrive in Pensacola, which means that soon complete financial responsibility will shift to the homeowners.

As I wander down Belmont, Romana, and Intendencia Streets I pass giant Bismarck palms and live oaks. I pass the Cajun Market with a big white U-Haul out front, a handwritten sign taped to the windshield. It reads, "Food Donations for Flood Victims in Baton Rouge." I walk past a sign for the Blue Wahoos Stadium and the New Tabernacle Baptist Church. Though I have just arrived in Pensacola, I decide to knock on the doors of a few people on the list.

The first place I stop, an elderly black woman opens the door but not the metal gate. When I tell her about my work and ask if she is planning to take out an NFIP policy of her own, she says, "My husband just had a stroke and they amputated his left leg. I'm on the way to the hospital to visit him now." Then she closes the door.

The second house where I stop, a white guy in a ball cap says he thinks about maintaining his policy but he and his wife don't have the funds to procure an elevation certificate, essential for a reduced rate. "We're in limbo," he says. "And we're on our way to visit my relatives." Then he too closes the door, turning the dead bolt behind him.

There is one more resident I want to speak with before heading to dinner. The name the paperwork gives is Robert Brown, the address 324 De Villiers.

The farther I walk down De Villiers, the more sparsely situated the small houses become. At one point I stop to ask a couple if I am going the right way. "I'm looking for number 324," I say.

The young man takes a sip from his Big Gulp. "Keep going," he responds.

Behind my back I can hear his girlfriend hiss, "But what could *she* want *down there?*"

On the surface this far end of the Tanyard looks like a depressed neighborhood in all the familiar ways—random pieces of furniture on the sidewalk, chain-link fences, broken windows—signs of basic human endurance coupled to long-term municipal negligence, discriminatory lending practices, and the stagnation of the minimum wage. But the closer I get to the bay, the more I begin to spot the telltale markers of flooding and flight. Many of the homes that sit on the ground bear watermarks—the paint is one color and then abruptly transforms into a lighter shade. Like half-dipped Easter eggs.

Finally I arrive at number 324, a baby-blue cottage propped up on cinder blocks. I knock on the door and step back, waiting. Then I knock again. I try the knob. Only then do I realize that the door is nailed shut. Tiny rounds of galvanized steel line the edges of the wooden frame.

I walk around the side of the house, where flood damage peels the blue paint from the laminate siding, and try the back door. The windows are broken, and beside me, on the stoop, sits a melted vinyl chair. Soot scars the backrest. Looking at this, I imagine that Robert took the little federal aid he was offered and, instead of rebuilding in place, fled. Maybe he even had a bonfire to mark the end of his former life. I squint into the sun that is about to set. Across Zarragossa Street stands another abandoned home. And just beyond it a third. This end of the Tanyard isn't just depressed; it's deserted.

Retreat from the coast and from flood risk might end up being primarily voluntary, I think, with those who can least afford to stay leaving first. But then my thinking morphs; it is incorrect to call flight from a blighted neighborhood—where the residents have been federally mandated to carry an insurance policy they cannot afford—voluntary. When Robert Brown left his home, that decision had in part been made for him. By the encroaching seas and the rainwater that sat in the streets, by the unwillingness or inability of the local government to update the stormwater infrastructure, by the federal government's need to get more at-risk homes into the flood insurance program in a last-ditch attempt to keep it from going under, by the simple math that making mortgage payments on an apparently uninhabitable home probably seemed like throwing good money after bad. If I am correct and Robert left of his own accord, then he left—unlike those in Oakwood—without the aid of relocation programs. Doubly failed by the municipal and federal governments.

The streetlights sputter on and the heady scent of angel trumpets fills the air as I walk back toward Palafox. If I had his number, I would call Robert and ask him what *exactly* made him leave the Tanyard. It occurs to me as I pass another flood-worn home, the transoms thick with vines, that perhaps the answer is simple and remarkably salient: people will leave only when they can no longer afford to stay. And, more importantly, the issue of affordability often slices only one way. In Gulf Breeze I met a couple who had used a mix of federal and personal funds to raise their McMansion high above the highest high tide. The county paid $124,000 and the couple provided a 10 percent match. The new version of their old home even came with an elevator to help ferry groceries up from the garage, where their BMW was parked. Meanwhile, in the Tanyard, I walk past a string of single-story cottages that have all been left behind. Battered mailboxes lie in front yards. Sediment from the last storm is papered over the lower portions of dozens of flood-ruined homes.

Now I am thinking about Alvin, who, for the time being, lives two blocks inland. He thinks he can prop his trailer up on cinder blocks. He thinks he can petition city hall to improve the storm-water infrastructure. But no matter what he does, every time high tide and a rainstorm overlap, the Tanyard will be full of water. His vulnerability is not something he alone can insure against, mitigate, or overcome. One day that water will warp his home into an unrecognizable shape, a place he doesn't have money enough to remodel. And then—assuming he lives through the storm—he will, like so many others before him, flee.

*

Eventually I tell Samuel that I cannot continue our professional relationship and I tell him why. First he says, "Oh my god." Then he says, "I had no idea." Followed by, "I don't remember." And then, "I had no further intentions." He says, "I love my family." And, "Let me know when you get over it." The words spill out of him fast like floodwater.

He can't stop talking, so I invent a student knocking on my door and hang up. I don't present at the National Academy of Sciences. I don't take the senior fellowship. I don't coauthor an article with him. When I put down my cell phone, I realize I have been shaking. It doesn't keep me from writing every trite word Samuel said on a succession of neon sticky notes that I immediately affix to the wall above my desk. Didn't I wish to somehow help him see what others could not; didn't I wish to introduce him to the most vulnerable living along our drowning coast; but didn't I flee, when I felt my body threatened?

For months, I fear I have failed Alvin and all of the other people I interviewed in the panhandle. That by walking away from Samuel I was also walking away from the occasion to get their voices heard. But the more I learn about risk, the more I know that fleeing is wise. It is something we ought to help each other do. In my case, it was easy enough. The immediate repercussions were few, aside from a handful of missed professional opportunities and a feeling of guilt at not having done more. My position and my privilege guaranteed my recovery. So many of those living along the damp fringes of our country do not share this sense of security. For now, when they flee from risk, it is less clear who or what will catch them.

I am done dreaming the earth undrowned; it is no longer a useful skill. Because what is happening in the Tanyard is happening to every denizen of the United States, every citizen of the world, even as it is affecting the most vulnerable among us first. James Baldwin wrote the following words about the murder of Emmett Till: "It's a terrible delusion to think that any part of this republic can be safe as long as ... members of it are as menaced as they are. The reality I am trying to get at is that the humanity of this submerged population is equal to the humanity of anyone else, equal to yours, equal to that of your child." "Submerged": Baldwin's metaphor has turned literal in the decades since he wrote it. His words remind me that our collective security will be arrived at, should it come at all, as a result of our ability to reckon with our country's history and how it has left so very many bodies unjustly exposed to risks that only continue to mount. "Guilt is a luxury that we can no longer afford," he continues. "I know you didn't do it, and I didn't do it either, but I am responsible for it because I am ... a citizen of this country and you are responsible for it, too, for the very same reason."

On Opportunity

Chris Brunet: Isle de Jean Charles, Louisiana

AS FAR AS CHANGES GO SINCE THE LAST TIME YOU WERE down here, well, for one, I'm bald. But really the biggest change is that it looks like the relocation of the community of the Isle de Jean Charles will happen. Through HUD [the Department of Housing and Urban Development] and the Lowlander Center, Albert, our tribal chief, got $48 million set aside to help us move. As you know, Albert has been pushing for relocation for the last sixteen years. All the other times, the place they wanted us to go was unacceptable. Or else there was not enough money to really do all that much. But finally with this try here it looks like it's going to work. With this grant we have more freedom and more choices, which is to our favor. The more choices you've got in any kind of change the better off you are.

The relocation project is going to be a community development. It's going to be the rebirth of the community of the

Isle de Jean Charles. But we don't know where just yet. We already had two meetings with HUD, where they called the island people together at the Montegut gym. They wanted to find out what it is that we wanted. I am so glad they did that because surprisingly so many good ideas came out of those meetings. Some brought up: What kind of drainage are we going to have? Some brought up: What kind of playgrounds are we going to have? We even brought up the idea of a fishing pond because we want to have the same kind of life over there that we have here. Mostly we want to be as close as possible to the island and not in the city.

Over the years, with the hurricanes and the land loss and the flooding, many people have been displaced. It got to the point that if something wasn't done then eventually there would be no Native community, no more people of the Isle de Jean Charles. Many of those that left, it looks like they're going to be included too, and I think for them especially this relocation can do some good. The island is already a skeleton of its former self and that's what's happening inside the community as well. When we relocate to higher ground we will at least be able to hold on to each other. I mean if we can stay together, then we won't have lost as much.

I've shared that thought with others, but saying it again to myself and to you right here, it is like, yeah, that makes sense. I mean really we are talking about having to choose to move away from our ancestral home. I know a lot of people figure we would be celebrating, to be moving to firmer ground and all. But it's not like I threw a party when I heard about the relocation. I'll be leaving a place that has been home to my family for right under two hundred years. I go all the way back to the island's namesake, Jean Charles Naquin.

Those of us out here are so tied into the Isle de Jean Charles. It's all we have known for the last eight generations. I spend most of my days right here. You know I am Choctaw, Native American. For us moving is not just about getting up and making a career move. We're actually leaving the place where we belong.

But what is the future of the island? Me, I know I could finish off my life on Jean Charles. I'm fifty-one years old. But then what about the next generation, those kids that are teenagers today? What about my niece and nephew, Juliette and Howard? I think about them at my age. I mean what is going to happen between now and then? In less than one hundred years, saltwater intrusion has taken away so much of the island. You are talking about an area that used to be eleven miles long and five miles wide. Now it's two and a half miles long and a quarter mile wide. So when I choose to be part of the relocation I'm making a decision today for tomorrow. I'm not making the decision out of desperation. I'm not making the decision out of fear. I'm just uncertain about what is to come. And I'm trying to take advantage of my chances while I have them.

I guess at the moment, I don't know what to say. I'm not having a great difficulty thinking about tomorrow today. It's not throwing me into a big confusion. But to arrive at this decision isn't easy. I don't know if I can offer any lesson to other people in the same or similar situations. Maybe the only thing you can learn from me would be to try to see your life so many years down the road, and if you see trouble, if you fear you're going to have too much against you, if the water is going to keep on rising, then it might be a good time to go.

Right here from where I am sitting down, I'm going to count off . . . one, two, three, four . . . five . . . six, seven, eight, nine, ten, eleven . . .

twelve, thirteen, fourteen … fifteen. Fifteen trees. Gone. You can't even see the stumps of some of them no more. Right there, right on the other side of the road. There was a time you could walk between these two big oak trees and the branches would cross over and the Spanish moss would hang down and it looked like you were under an archway.

You could say once the salt started to get into the water table, that was when everything started to die. But it was a gradual process. Now, today, we see we done lost so much. We can only say that in the present though, because back when it was starting we really didn't know. It has got me thinking: What is the definition of the coast? What is coastal Louisiana? If everything is shut down here, what is Louisiana doing being the "Bayou State"? What is it that makes Louisiana the sportsman's paradise? Where is it that people will go whenever they leave up north to come down here and see the beauty of Louisiana?

I am sitting here, a coastal resident surrounded by water and coastal erosion. I am moving in. I know for myself that no matter what kind of technology they possess they cannot bring it back to what it was. There is no way to recreate Louisiana's coast, all the bayous and lakes, all the shrimping and the crabbing, and the other animals that lived out here. But if something can be done to slow down the tide and they can save what they still have, they should do it. If you come right down to it, we are all up against coastal erosion. We are all impacted by it. It has taken the island away from us, and in the future it is going to take other places as well. I may be uncertain about a lot of things, but not about this here.

Goodbye Cloud Reflections in the Bay

Isle de Jean Charles, Louisiana

FOR A LONG TIME, I THOUGHT THE STORY OF THE ISLE DE Jean Charles ended beneath Edison's persimmon tree. With him and Chris both staying in the place that shaped them, even when that place was changing irrevocably. But in May 2016, nearly three years after my first visit to the island, the *New York Times* runs a front-page article with the headline "Resettling the First American 'Climate Refugees.'" Above the fold is an aerial photograph of what little remains of the island.

I call Chris immediately and ask if he is going to leave.

"Yes," he says. "I'm not celebrating, but I'm going."

Three months later, I am back on the island for what I imagine will be the last time. As I drive out on the Island Road I try to notice what, if anything, has changed. Perhaps the oil refinery is sitting a little lower on its stilts—or perhaps a few more cracked cypresses have dropped into the encroaching sea? Though I know there must be *slightly* more water here, to my eye everything is just as it was the last time I came to Jean Charles. Still my knowledge that relocation will go forward alters my relationship to the landscape, if not the physical land itself. I feel as though I must say goodbye to every little thing I pass.

Goodbye sno-ball shack and snowy egret.

Goodbye sign that says, "Home Lift Slabs or Pier."

Goodbye cloud reflections in the bay.

Goodbye algae-covered jetty, and the men with their taut lines and their pickups parked alongside.

Goodbye shrimp lurking in the seagrass stalks.

Goodbye persimmon tree.

By the time I reach Edison's moss-green cottage I am getting pretty good at saying goodbye. Inside, his wife, Elizabeth, is watching television. At first she doesn't want to tell me where Edison is. Then I explain that I interviewed him some years ago and that I have brought a small gift. Eventually she acquiesces. "He's down on the bayou, the one across the road," she says, "throwing the cast net."

I walk down the blowsy bayou spine and there in the first cove is Edison, his shirt sun bleached and soaked through with sweat. A pile of oyster shells heaped at his feet. He is heavier than before, his hair longer.

"Edison," I say, "I'm not sure if you remember me. I was here a couple years ago, and—" I am uncertain of how to continue, of how to distinguish myself from the parade of journalists and writers and documentarians who have probably been here since. "You gave me a cluster of oyster shells from your altar, and I brought you something in return." I hand him a hand-painted ceramic rooster from Providence. "They say it brings good luck into a house," I add. He motions for me to lay the figurine in the high grass, his hands salty and wet with bayou brine.

"How's the altar?" I ask.

"Gone, flood took it all away," Edison says.

I look back across the road to where the altar once stood and see nothing but layer upon layer of roseau cane and cordgrass. I turn back toward Edison, thinking through my list of questions. They are all different versions of the same thing: Are you going to leave the island?

Just then two representatives from Concordia, the architecture firm hired by HUD to orchestrate the resettlement, appear. They have mod sunglasses and accents from away. The older of the two looks at me and asks if she is interrupting something.

"No," I say, curious to see what they want.

Edison shifts his weight back and forth—momentarily, it seems, feeling surrounded.

"We're conducting a survey," says the younger one.

"I'm not leaving," Edison responds. "This is my home. You can write that down."

"What is it that you like about your home?" She flips open her clipboard.

"I have no interest in moving into a poorly made house fifty miles inland," Edison says. "If you want me to move, why don't you let me find my place, where I want? I could build in my son's yard in Bourg."

"In order to be part of the resettlement you have to move in with the rest of the community."

The older woman removes her sunglasses and adds, "When the relocation is complete, the road to the island won't be repaired after a storm."

Edison looks right at her, unblinking. "I know you're not saying everything you know. I can tell by your eyes," he responds.

To which the woman says nothing at all.

When I first heard that nearly $50 million had been allocated to the islanders I was ecstatic. In the three years since my initial visit to Jean Charles I had watched a handful of other coastal communities band together to try to secure state, federal, and private funds for their retreat. Some, like the residents of Oakwood Beach, had been successful, while others, like those living in Kivalina, Alaska, had not. Then came the National Disaster Resilience Competition, which grew out of a series of planning workshops hosted by the Rockefeller Foundation in Sandy's wake.

The goal was to encourage communities to consider not only how they could recover but also how to avoid future losses. All state and local governments with major disasters declared between 2011 and 2013 were invited to participate and submit a design proposal that would use recovery funds in an innovative and forward-thinking manner. HUD awarded over $1 billion to thirteen different project proposals. Louisiana received nearly $100 million to protect coastal wetlands, retrofit communities to withstand increased flood risk, and, in the case of the Isle de Jean Charles, experiment with relocating residents *in advance* of the next storm. The state would purchase open space inland and build dignified residences for the islanders, likely even those who had left long before the most recent flood. From a remove, it seems like a win-win.

But up close, relocation is more complicated than I had imagined. The woman's comment about a future where the roadway won't be repaired acts like a threat on Edison. He digs his heels in even deeper.

"Don't come in here and take me and put me where you want," Edison says, his voice getting louder with each word. "What am I going to do inland, watch TV and get old quick?" He has a point, I think. I wouldn't want anyone telling me where I had to live, especially when I had figured out a way to be happy where I was.

"There's talk of building a pond on the project site so you can keep fishing."

"You call casting a line into a stocked puddle fishing?" Edison snaps.

"We hear you, loud and clear," says the younger of the two. Then the pair turn around and walk away. If Edison won't fill out their questionnaire, they aren't interested in keeping the conversation going.

He looks at me then as if to say, *You're still here?*

I tell him about my work on this book, about the inadvertently coercive retreat in Pensacola and about Oakwood and the buyouts there, how residents got to choose where they wanted to go. But while Oakwood was full of working-class whites, postal workers, and civil servants, those who live here are seen, first and foremost, as Native Americans, even if the federal government doesn't formally recognize them as such. As I speak I begin to see that the communities with the least options going into a flood have predictably fewer paths toward equitable relocation afterward.

"I guess one of the reasons the grant money was allocated to the island is because they want to bring the tribe back together," I say. It sounds like I am trying to convince Edison to leave, so I add, "You know, Albert says, 'If you put the pieces of a chopped-up snake alongside one another it will reanimate. The snake will become whole once again.'"

"If we do have to move, make sure the ones in charge walk out front," Edison says. I nod and a bead of sweat falls into my eye.

Together we stand in stunned, hot silence. The air is full of lovebugs. I watch two land on Edison's faded blue T-shirt, right next to the dime-size hole worn through near his belly button. After mating, adult pairs remain coupled for days, even in flight. Their larvae feed on decaying vegetation, which is likely why the number of them living in the Gulf Coast's drowning wetlands has skyrocketed over the past fifty years.

"And the shrimping, how's that?" I ask, changing the subject.

"Better than it was." Edison leans down and tilts the white five-gallon bucket toward me. It is at least a third of the way full, hundreds of live shrimp squirming at the bottom. For a while we talk about the tides and the return of the Louisiana browns after the disastrous Deepwater Horizon explosion five years earlier. "Come back later and I'll give you some, cleaned and all."

I tell Edison thank you, that it is good to see him and that I will come back closer to sunset, when it is cooler. I am about to step over the rooster and go when I turn to him and say, "I've put it there on the ground."

"I won't forget it. Not ever," Edison responds.

I suspect that most who travel to Jean Charles these days either want to carry the story of the island away from the island, like me, or else are trying to convince Edison to move. Few, I imagine leave something, anything, behind.

On my way back to the car I pass a new sign that Edison has tacked next to the old one. It reads, "THEY SAY THE ISLAND is FADiNg AWAY SOON WE WiLL NOT hAVE A ISLANd LEFT. iF THE ISLAND is NOT GOOD STAY AWAY. MAY God BleSS THE ISLAND!!!"

✷

I park my car in the empty lot across from Edison's place and walk the rest of the way back out to Chris's house.

I walk past the husk of the home where I once saw Howard and Juliette jumping on an abandoned mattress. The roof is missing and the pilings are all charred. Past the empty fire station and the abandoned fishing camps. Past trailer homes with their contents spilling out like organs from a wrecked body.

Eventually I arrive at Chris's. I walk up the long flight of stairs to the front door. Everything looks exactly as it was, except that the swimming pool is gone and the sheets in the interior have all been taken down. Chris is there, sitting behind a hot plate, frying hamburgers. He hands me a slender patty sandwiched between two slices of Wonder Bread and I set it on the table so we can embrace. Ever since my first visit to the island, we have stayed in touch. Chris always calls on Christmas and Easter. I send him photocopies of the articles I write.

"Thank you," I say.

"For what?" Chris responds. "The hamburger? No, it's nothing. You have to eat, shore up, to make it through the day."

"No, thank you for taking care of me. When I first came out here, I was in the middle of leaving a life I'd built and you fed me and welcomed me into your home. Showed me a kindness I'll never forget."

"Aw, child, all of us out here remember you. I'm glad you came out the other side, that's all."

I hand Chris a ceramic rooster like the one I gave to Edison, a hand-painted heart on its chest. "There are a lot of Portuguese in Providence, the town where I moved," I tell him. "Over a hundred years ago, they came to Rhode Island for the fishery jobs, and many of them stayed. They say these roosters bring blessings into a home, no matter where it is." He unwraps the animal from its plastic and places it right in front of his big-screen TV.

"Let's go down to the porch," Chris says. "There's a nice breeze."

We spend the entire morning beneath the house, following the shade as it moves from the western side of the slab to directly underneath. I tell Chris about Rhode Island, my new job and new beau. He tells me about Juliette's sudden transformation into a teenager and Howard's football practices, touching here and there on the relocation.

"Ever since the *New York Times* ran that article, there have been at least three television and movie crews out here every week. Since I don't tend to go far," he says, tapping the wheel of his chair, "it's kind of nice. The world coming to me and all. Last week there was a group from Holland." I look at Chris and smile. In case I am jealous, he adds, "But I don't keep in touch with anyone else, that's only you and me."

Just after noon a white Cadillac Escalade parks in the driveway and a television crew pours out, unloading camera equipment, boom mics, and rolling dollies. They don't come over.

"I forgot about them," Chris tells me, lowering his voice. "I *think* they're from *National Geographic*."

I tell Chris I don't want to keep him. I will be sure to return the next day, to say goodbye. I step back and watch the young men with their grizzled beards, tortoiseshell glasses, and ball caps swarm around my friend. First they ask him to roll over to the eastern side of the slab where the sun isn't so bright. Then they hold up light meters and hand Chris a lavalier mic. They prompt him to say something for practice.

"I am making a choice today for tomorrow," Chris says. The words flow out with rehearsed ease.

I walk back down the length of the island to my rental car. When Edison sees me coming he descends the set of stairs that connect his cottage to the road. At the bottom he hands me a ziplock bag full of freshly caught shrimp and a second of shrimp jerky.

"I dried it, spiced it myself," he says.

That night I sauté the fresh ones, drowning the whole lot in butter and mixing it with spaghetti. It is the best scampi I have ever had. The flavor deepened by my knowledge that Edison caught these shrimp himself, in the bayou behind his house. Though I hate to admit it, I think, *Someday that bayou will be gone.*

"Goodbye," I say aloud to no one, popping the last pink shrimp into my mouth.

✸

My visit to the island is short—I've squeezed it between Pensacola and the start of the new school year—so the next afternoon I drive out to Chris's one last time. Together he and I and a cousin of his, Walton, drink a dozen cans of Bud Light. Unlike my previous work trips, this feels less like research and more like a farewell or a homecoming; I can't exactly tell which. I have a hard time separating excavation from elegy. The confusion of the two, however, leads me to conclude that drinking weak beer under Chris's house is a perfectly acceptable way to spend my final afternoon on Jean Charles.

Chris tells me that the wife of his other cousin, Dalton, gave birth to a baby boy in the night. From there the conversation turns nostalgic, with both men remembering their fathers when they passed. Walton's had contracted a debilitating form of diabetes that took his right leg and later his life. Shortly after he died, Gustav wrecked the family trailer and Walton moved to Houma. Still within spitting distance of the island, but far enough in to avoid all but the strongest of storms.

"I make it out once a month, sometimes more," he says, walking over to his pickup truck to pull another can from the back.

Walton and I talk about all the different drivers behind the National Disaster Resilience Competition, speculating on just how much of the allocated funds will end up aiding directly in the relocation. He thinks that everyone from New Orleans to Houma is going to want a slice of the pie and that far less than anyone imagines will end up making it all the way out here. Growing tired of the steady stream of acronyms we use to describe the various stakeholders, Chris eventually interjects. "Quit using all those fancy words," he says, slapping his hand on the table and taking another sip of cold beer.

It's a joke, and I can tell that it is not designed to sting. Chris makes fun of just about everyone he loves. But it makes me think that no matter how close we have become over the years, I will always be from away. No matter how good I get at leaving—a loved one, a rental apartment, an adopted state, an advantageous work relationship—I am not an islander, not the one who has to figure out how to say goodbye to a swollen home on a sodden island that for a good two hundred years few cared for or visited.

Chris pulls out his photo book. He has got it well organized now, with clippings of his favorite articles about the discovery of oil on the bayou laminated alongside Polaroids of past storms and portraits of each of his family members.

"When folks ask me what the island used to be like, I show them this," he says with pride.

After three Bud Lights I am both bloated and buzzed. The early evening light is a clear and glowing thing and I am deeply glad I made the drive. Regardless of how familiar I have become with the idea of human communities moving in, looking past the bare arms of the dead cypress trees that line Chris's drive out into the open water that surrounds his soon-to-be-former home unhinges me still.

"What are you going to do with your pirogue?" I ask, pointing to the flat-bottomed boat near the refrigerator.

"Don't get out into it often now," he says. "Last time was about three years ago. And I tell you what, when I looked back at the island, most of it had disappeared. I knew then that at some point in the future Jean Charles would be gone." Like Yellow Cotton Bay and English Bay in nearby Plaquemines Parish, its name will eventually be removed from the maps. The inked black letters will fade into blue. After the last crumb of red earth that once was Jean Charles dissolves into the Gulf, after the emergency that is the loss of a home place passes, something else will emerge from the salt water. The story of the island will replace the island itself, and those who loved Jean Charles will guard that tale, just as the Penobscot in Maine still celebrate the caribou a century after colonists killed the very last one.

✻

Months later I am asked to respond to a lecture given by Harriet Tregoning, the former director of the Office of Economic Resilience at HUD and the woman who launched the competition paying for the relocation of the islanders. The event organizer hopes that I will add an "on-the-ground perspective," which I understand as an invitation to share Chris's and Edison's experiences with the planners, bureaucrats, and environmental engineers whose work affects those living at the hard edge of climate change. It felt like a huge responsibility, but one that I was honored to have been asked to bear.

During Harriet's lecture, she shows only two slides of the island. The first is a then-and-now comparison. In one aerial photograph the island is huge and green and pastoral, and in the other it is nothing more than a solitary road surrounded by the Gulf. "This is how much land has been lost in less than fifty years," Harriet says. The second slide is of the Island Road in a storm. Water laps over the crumbing edges of the macadam while a single lone figure forges on, out toward the shell of Jean Charles.

I realize instantly that the image is a still taken from the movie *Beasts of the Southern Wild*. The figure isn't an actual resident of the island but the film's six-year-old protagonist, Hushpuppy. Her afro is backlit in the storm light and her tattered white

T-shirt slips from her right shoulder, weighted down with rain-water. She's walking back out to the Isle de Charles Doucet (as it is called in the film) because she, like the rest of the residents, refuses to leave. It is a shot that Benh Zeitlin, the director, hung around deep into hurricane season to get. While the person in the image is fictional, the landscape and the flooding depicted are not. Likely the scene was filmed during Tropical Storm Lee, when the tidal surge topped five feet.

I wonder if Harriet knows that her image was staged; if she knows that in another context it signifies something else en-tirely—resiliency in place—and the extent to which its origins matter. The funny thing is, when Benh Zeitlin made his magi-cal-realist indie epic he couldn't have known just how funda-mentally it would end up mirroring real life. He didn't know that millions would be set aside to get the islanders to move in, and that there would be some, like Hushpuppy, who refused to leave no matter what. If nothing else, when you understand the prove-nance of the image illuminated above Harriet's head you can see, clearly, that the environmental apocalypse we often think of as existing only in films is already with us. The lines between our imagined futures and the present tense grow increasingly blurry with every passing day.

In the Q&A session a man in a blue checkered shirt moves toward the microphone and asks the million-dollar question: "If retreat is a useful adaptation strategy, how can we make more people interested in giving up their homes?"

"I'm not sure you can," I hear myself say. "It has to be a deci-sion a person arrives at themselves. Those who feel forced aren't going." This logic holds true for most anyone, but especially for those whose homes were hard won and left alone by the greater community as they came undone.

Where Chris sees opportunity, Edison sees coercion. Where

Chris has made Jean Charles his own through a historian's lens—by the picture of the island, past and present, that his photos and memories paint—Edison knows that his identity is tied to a very particular physical place. Where Chris feels a profound connection to his Native community, even should the setting change, Edison senses that in leaving he would lose perhaps the largest part of himself, the part that is of the island.

On the one hand, I want to believe that learning to retreat is simple enough, with concrete steps and consistent outcomes. That it means inviting those who are relocating to tell the story of their own departure. That it means financing the move. That it means changing flood insurance policy so you can take your claim money and, instead of rebuilding in place, start fresh somewhere else. That it means making sure it is not just the poor who are expected to leave. And yet, on the other hand, I respect the connection Edison feels with Jean Charles. He doesn't simply live on the island; he knows that who he is, his very sense of self, is linked to the land where his entire life has taken place.

When I reflect on my time with both these men I realize that there is, in the end, something elemental that they share. Both have the ability to read their surroundings and respond; they align the stories they tell with the decisions they make, even when those decisions differ dramatically. Stay or go? Retreat or remain? In the face of so much change, both Chris and Edison are retaining their control—if not over the physical world, then over the words they use to make sense of their experience in it. The longer I spend on our disintegrating shoreline, the more this strikes me as an adaptive technique that humans alone might have.

When people love a place, it can change in shape and we can adapt our love to its transformed state. We can make do with less. Catch fewer shrimp. Sink cucumber seeds into soil we have placed in a repurposed bathtub. Plant persimmon trees. Or else we can pull up our roots and move in. When we suffer unthinkable losses we can conjure images of what once was. In the future, the children of Howard and of Juliette, should they have them, will hear thousands of stories of the island, some real, some transfigured through a game of telephone that began even before their grandfather was born. I imagine that when Chris's descendants are living high above the highest tides they will say, "Long ago the island was a magical place. So magical that many who once lived *there* moved *here* to keep its memory alive."

PART III

Rising

Connecting the Dots

H. J. Andrews Experimental Forest, Oregon

THE RUFOUS HUMMINGBIRD IS NO LARGER THAN A SPOOL OF thread. The mass of the average adult equals that of one and a half pennies. Females are bigger than males, and both eat three times their body weight most days to stay alive. Despite their size, these tiny rust-throated birds travel five thousand miles each year between their wintering grounds in Mexico and the Gulf Coast and their breeding grounds in the center of the Pacific Northwest. Measured in body lengths, their migration is the longest in the world. The oldest begin the journey in late January. The youngest follow shortly thereafter.

When I show up at the H. J. Andrews Experimental Forest in late May the rufous have already arrived, plunging their feathery tongues deep into the first alpine wildflowers. I too have traveled far to get to these fifteen thousand acres, nestled in the luxurious folds of Oregon's Central Cascades. But unlike the humming-birds', my journey wasn't nectar fueled. I watched two movies and ate a ham sandwich while my Boeing 737 burned through thousands of gallons of jet fuel. I do, however, walk the final twenty feet to a cabin in the woods on my own. This will be my home for the next two weeks while I serve as the forest's writer-in-residence.

On my third morning, it is a little after eight o'clock and I am outside; the clicking of red tree voles high up in the surrounding old growth rinses my mind clean. I have been awake and writing since before dawn, and I needed a break. I needed to feel the last of the morning mist latch on to my hair. I needed to stare dumbly at the deep rivulets and landscapes in miniature running up the trunks of the nearby Douglas fir trees, which have been thriving in this little grove for well over four hundred years. This easy proximity to a sliver of the natural world that resonates on a time scale so utterly different from my own: that is why I am here. I am hoping these weeks awaken something in me, something in the language that I use to describe events so large they resist my pinning them into the pages of this book.

I walk past the research scientists' bunkhouses. They are low slung and the color of mud. Hand-carved wooden signs hang above the entryways, labeling the buildings with names linked to local geographic features and species. First I pass Quartz Creek, then the Rainbow Building, named after the beloved endemic trout. When I reach the balcony of Roswell Ridge, where the bird crew sleeps, I stop. Someone has lined the railing with hummingbird feeders. There I spot my first rufous: pouring its slender beak into a glass spigot, sucking up the syrup below.

If these slight birds spun out silk as spiders do, each one would run through 8,849 spools a year in its migrations. When the rufous flies away, I imagine a single iridescent string trailing behind its feathered body. And then I imagine the many thousands of others in the Andrews just like it. If the continent is a quilt, then these hummingbirds—so much smaller than my own hands—place the stitches that hold the fabric together. I squint and try to distinguish one wingbeat from the next as another rufous approaches. What extraordinary creatures, I think, weaving here and there—mountains and lowlands—together with their windblown little bodies.

＊

The H. J. Andrews Experimental Forest is one of twenty-eight Long-Term Ecological Research (LTER) centers scattered across the United States. In the early eighties, when the National Science Foundation realized that almost every project it funded was driven by individual researchers and limited in scope—often spanning no more than five years, thanks, in part, to the steady tick of the tenure clock—it began to invest in a series of outposts dedicated to the study of ecological processes that play out over much longer periods. While the idea seemed novel and perhaps indulgent thirty years ago, today these research centers are generating some of the most relevant climate change data in the country. At the Bonanza Creek station in Alaska, they track the impact of arctic ice loss on human and nonhuman travel across remote landscapes; scientists at W. K. Kellogg Biological in Michigan measure the rate of CO_2 respired by different agricultural crops; and just a couple hundred miles south, at Konza Prairie in Kansas, they are studying the impact of increased rainfall and warmer weather on the productivity of grassland ecosystems.

Along with the handful of climate studies that the Andrews hosts, it is one of the few LTER stations to devote funding to the arts. Its Long-Term Ecological Reflections program invites two writers annually to take up residence in the forest, to visit and record their responses to a predetermined set of study sites. Each writer reflects upon the exact same spots as those who went before, and collectively they generate a creative record of the changing relationship between people and forests over time. When I accepted the invitation in 2016, the project was already in its thirteenth year out of a planned two hundred. I had, completely independent of the residency, read many of the essays produced during the first decade of the program, so I thought I knew what my time in the forest would yield.

I thought I would spend my mornings revising this book, then hike madly through the afternoons, making sure to visit each of the four reflection plots littered throughout the Andrews. At night, I imagined, I would cook myself coho salmon for dinner and fall asleep to the sound of Lookout Creek tumbling over the downed bodies of five-hundred-year-old trees.

But when I spot that first rufous, my expectations about what the residency will produce begin to shift. Not revision, but expansion. Even though I am 1,350 feet above sea level and a good hundred miles east of the nearest saltwater marsh, I cannot stop thinking about the changes taking place closer to the sea. When I look at the iridescent feathers twitching on the rufous's belly, I do not see the pump and flow of blood beneath the skin or the stitching of flesh atop its wind-shot bones. I do not see a bird exactly. Instead I see a map of its migratory route, and the many swamps and wooded lowlands that it passes through along the way.

*

About a week in, I meet Sarah Frey, one of the few researchers at the Andrews devoted to studying the impact of climate change on the forest. She is wearing a green down parka and a headlamp. It is nine o'clock in the morning when I reach the open field behind the Rainbow Building, but Sarah has been sitting there since dawn, capturing hummingbirds and implanting them with tiny tracking devices called Passive Integrated Transponder (PIT) tags. She is exactly what you might expect from an early-career female scientist: straight faced, sleep deprived, and curious.

In 2008 she and her husband, Adam Hadley, set up a new three-part long-term study at the Andrews. Step one: distribute strings of vertical temperature sensors throughout the forest. These generate a data set exhibiting, at an unprecedentedly fine scale, the range of undercanopy air temperatures across the mountainous terrain. Step two: track bird movement through the forest during breeding season. And step three: analyze how temperature is influencing bird distribution in the Andrews. In an age when climate change is expected to have pervasive negative effects on biodiversity, the husband-and-wife team wanted to know whether there were unexpected pockets of cooler temperatures scattered through the Andrews, and if birds had begun to seek them out.

"We've got one," Sarah calls to Adam, who is doing "surveillance" at the other capture net. He comes over to the folding table where Sarah is about to perform a minor surgery, pulls at his LTER baseball cap, and offers to narrate so his wife can concentrate.

She grabs the tiny bird the way a child grabs a thick marker, pressing all five fingers into its round torso so she can hold it up to my eye level. I am immediately impressed by the rufous's aliveness, the animal quality of its face. I see the features of my cat and also those of my lover reflected there—they are the only other beings I have recently inspected at such close range. And then there is the hummingbird's beak, which is far more menacing than I had expected. I see it and think: tusk on a narwhal, long and sharp and strangely disproportionate in size to the animal's body.

I pull closer, and Sarah reflexively loosens her grip. Sensing the possibility of freedom, the rufous darts into the dense undergrowth. "I really wanted you to get a good look and I lost the thing," she says, turning from me, disappointed.

According to the Audubon Society, by 2080 this glittering hummingbird will lose 100 percent of its nonbreeding range in the United States, most of which rests in the forested lowlands around the Gulf of Mexico, the place where cypress stands meet marsh. I think back to the hundreds of dead trees that line the Isle de Jean Charles's drowning bayous and know that the loss of this bird's winter range has already begun.

It might seem like a stretch to say that *here* is connected to *there*, and that the bodies of these small birds do the connecting. However, just as the Neapolitan immigrant brings a bit of Italy to New York City, and just as Colombians from Medellín carry the central highlands to the northern corner of Providence, so the rufous transport some piece of all the places they pass through here, to what Sarah calls the belly button of the Willamette National Forest.

Another rufous wings by in a blur, and I begin to wonder what will happen when their wintering grounds disappear. If the rufous don't have a place to live come January, will they return to the Andrews in May? If not, they won't thrust their long slender bills deep into the scarlet gilias and western columbines for nectar. They won't withdraw faces and bodies dusted in pollen. I imagine that if they don't do this, perhaps fertilization will become less likely. Fewer seeds will form. And if fewer seeds form, fewer flowers will reproduce, their Technicolor slowly seeping from the high spine of Carpenter Mountain.

At least that is one version of what could happen. It's also possible that another pollinator could rush in to fill the space left by the rufous. Or that the rufous could change its wintering grounds and its migratory route and end up drinking the nectar of other flowers on other mountainsides. I don't know what bothers me more, the possible loss of the rufous and all the biodiversity that partially depends on it, or that I have no idea how any of this will play out in the end.

*

A couple of days later, when the first of three record-breaking heat waves that will descend upon the state this summer has finally broken, I drive to the top of the Andrews, where the Carpenter Mountain Trail begins. It is the highest in the reserve, and even the parking spot has a view. The edges of the narrow path are littered with trillium and Indian paintbrush, with globe-flowers and Queen Anne's lace. Before I lived in Rhode Island, before Brooklyn and Southeast Asia, I lived here in the Pacific Northwest. Often I tell folks, "I miss Oregon like a person." And even though a decade has passed since I last relished the long and deliciously damp Cascadia winter—perfect for mornings at my writing desk, afternoons at Powell's, and weekends in the woods on raincoat-mandatory hikes—returning always feels a little like coming home. Every other year I make a pilgrimage to this moss-heavy range. It reminds me of the relationship many of those who left Jean Charles still maintain with the island. Though Walton doesn't live out on the bayou anymore, he returns at least once a month to throw the cast net and catch up with Chris.

Switchback after switchback, I work my way up. My mind empties as clouds begin to cluster around the mountains across the valley. Eventually the trail levels out, meandering through a stand of pistol-butted western hemlock encased in wisps of witch's hair, spectral green lichen, that sway in the thin air. One tree has fallen across the path. Up close the lichens covering its long body are unruly, each rubbery strand a neuron firing, each cluster an entire neural network. In winter deer and caribou feast on this stuff. Today the ends of many of the strands sport tiny saucers. My field guide tells me that these are rare and signal reproduction.

The witch's hair is not the only living thing here in the midst of making more. The Andrews serves as a breeding ground for hundreds of different species of migratory birds: ospreys, Pacific-slope flycatchers, black-throated gray warblers, nighthawks, Swainson's thrushes, lark sparrows, golden-crowned kinglets, and many, many others. Like the rufous, they are hardwired to take advantage of the intense productivity of northern ecosystems during the long days surrounding the solstice. And like the rufous, these birds travel across a transcontinental web of marshes, estuaries, wetlands, and bogs, a patchwork of feeding grounds and places to pause when exhausted by so much motion. In the early 1920s, the federal ornithologist Frederick Lincoln introduced the concept of the flyway, a migratory bird superhighway. Like us human beings, these feathered fellows are creatures of habit. When they travel they often retrace their routes, making pit stops in the same exact locations year after year. Today there are hundreds of refuges strung, like beads on a necklace, along the United States' four primary flyways.

Earlier that morning, after spotting a pair of ospreys turning slow and weighty circles above my cabin, I read the *Audubon Climate Report*, a guide to the future of North American birds. I was not surprised to learn that 40 percent of all avian species are migratory. In North America, a third of these nomadic creatures are considered at risk of extinction, thanks in no small part to the threat sea level rise poses to coastal wetlands. As I hike I think back to those findings and wonder if in the future we will ask ourselves the chicken-or-the-egg question in reverse: which went first, the tidelands or the osprey? The columbine or the rufous?

Soon the switchbacks begin to steepen, the moss to wither, and the rufous tracking stations to multiply. But I don't see a single hummingbird. Is it coincidence or a sign? The thought haunts the rest of my ascent. Species extinction, of course, means not individual deaths but the irreversible disappearance of an entire kind of animal. And yet in the uncanny way that even the most horrific events can become normal when encountered with some regularity, the word *extinction* no longer registers as astonishing. I realize on that mountainside that I have become deeply habituated to the thought of losing much.

Eventually the forest gives way to the summit, a fist of high alpine basalt that until just a few weeks ago was under snow. At its top sits a boarded-up fire lookout. It is too early in the season for the cabin's keeper to be here. I climb onto the wooden porch and the Three Sisters, with their volcanic, still-snowy peaks, swim into view. Cold air glides out of a nearby gully, making me pull my jacket from my pack. I eat three squares of chocolate wrapped in gold tinfoil. From far in the valley below comes the looping call of a solitary bird attempting to attract a mate. In my notebook I scribble, "I come to these mountains with my big questions, the way I imagine my grandparents went to god." Down in the valley, the bird repeats its call again and again. I crane my neck, turning my ear toward the wind-weighted Douglas firs, but hear no response.

*

There are 183 black dots on the laminated map of the Andrews that Bryan Doyle lays on the hood of my rental car. Each of the dots represents one of the sites where, six times in each breeding season over the past eight years, members of Sarah Frey's field crew—including Bryan—have listened for ten minutes and recorded birdsong in the canopy. This data will be mapped onto the vertically integrated temperature readings she has been collecting.

When the study is complete Sarah will have an illustration of how heat, and therefore changes in climate, influence the movement of breeding birds throughout the Andrews.

Of the 183 dots, points 1 to 182 are numbered sequentially. Then there is point 400, the northeasternmost spot in the Andrews's long-term bird count. It sits well over a mile from the nearest trail or road. It is 3:45 in the morning, and Bryan and I are about to attempt to hike to it in the darkness before the dawn.

I drum my finger on the map one last time and flick off my headlamp before getting into the car. Bryan settles into the passenger seat, brushing aside his two-foot-long beard, which is the ochre color of decomposing Douglas firs. I drive us deep into the Andrews, gliding through long tunnels of trees and salmonberry vines, gravel thundering on the underside of the car. We skirt the edges of high mountain meadows, roll through fields of thick fog that few will see. After forty minutes Bryan flips on his GPS and coolly says, "Here." I park the car in the middle of the road. There is no shoulder and no one coming up behind us.

Bryan puts on a baseball cap and I do the same. He hands me a walkie-talkie the size of a brick. "Set it to channel three," he says. "Then hit 'Talk.'" The air fills with crackling static. "Just in case," he adds, a phrase that I repeat.

Then he turns, steps off the unpaved road, and slips out of sight into a dense thicket of vine maples. Vine maple is related to the bulkier and more common broad-leafed maple, but it has dozens of smaller trunks instead of a large main one. On steep slopes like this, the trunks bend sideways, creating an interconnected web of branches like outstretched, grasping fingers.

"It's a bit like going through baleen," Bryan says from ten feet in front of me. He lifts a foot waist high and kicks it forward, navigating up and over a tangle of trunks. He totters in a controlled fall and the branches close immediately behind him.

When I try to imitate Bryan, my thigh gets caught, though inertia wants me to keep moving. Eventually I wrest my leg free. I begin to let gravity do more of the work and soon I too am half falling down the hill, slipping into the whale's mouth. Our world becomes darker as the forest canopy closes, the hill always against our backs.

"If you think this is hard, try climbing out," Bryan calls.

Even with my headlamp on I can't really see him. Instead I navigate by ear and by touch, letting go of the idea that I need to follow a trail to move through the woods. Twigs bend and pop underfoot. Perhaps this is what the future will be like, I think, as the places we have long navigated by disappear beneath the surface of the sea. At first the disorientation is uncomfortable, but slowly my mind and body unclench, embracing the unknown.

After thirty minutes the vine maples open up into what was once a Douglas fir plantation. We flick off our headlamps as the sky turns the color of whale oil, smudged and gray. The forest floor is dry and bouncy beneath our feet, dropping away and then rising back up in one long undulation. By the time we reach point 400, over an hour has passed since we started walking. My skin sticks to my clothing, but I resist the urge to peel off my jacket. When crawling through thickets of devil's club I need the extra layer. The shrub is so prickly that the Salish people equate its diabolical spines with the power to protect, and mix its ashes with bear grease to make ceremonial face paint.

"I had no idea birders were so badass," I tell Bryan, panting and throwing my pack on the ground.

"We definitely see more of the Andrews than any other crew."

Bryan holds up his watch to signal that he is about to start the count, then cocks his head to one side, steadies his clipboard, and starts taking notes. He recently passed a series of tests devised by Sarah to measure competency in birdsong identification. His colleagues, who have not yet done the same, must lug a recorder the size of a bread box through the backcountry. Bryan relies on his ear alone. A song cascades down from the canopy like water tumbling over polished stones. From a bit farther up the hill comes the steady trill of a miniature machine gun firing. Some calls are straight chirps and others seem to move back and forth, like a conversation between old friends.

When the ten minutes are up, I ask Bryan what he has heard. There are three Swainson's thrushes in from the Andes. A Pacific wren visiting from Nevada and other states east. A Steller's jay and a varied thrush, both locals. A Pacific-slope flycatcher up from the mountains of Mexico. And a golden-crowned kinglet who, like Bryan, came from the Midwest.

As he describes their many different points of origin, it occurs to me that most of the birds and both of the humans in this stand of Douglas firs are in from out of town. Summer visitors, vagrants, migrants, travelers all. Each of us will leave the Andrews and its old growth when the days get long and rainy. And each of us will likely return again. Perhaps in this way Bryan and I are not that different from the rufous and the other birds overhead. We too fly in an endless loop. We are snakes with our own tails in our mouths rolling downhill. There is no difference between the beginning and the end. No singular home, pulling harder than the rest.

In the winter, when the bird count season has closed, Bryan works the graveyard shift at a liquor store in Saint Paul, Minnesota. He also fronts a one-man band called Typewriter. His second EP, which he plans to release this summer, is going to be called *Songs from Space Station H*, meaning the H. J. Andrews Experimental Forest.

"This place is so removed from everything else. In this weird way, being out here, you feel sort of out of time. And yet, you know, it is so keyed in as well . . . ," Bryan says, trailing off, his attention turning to the granola bar he has just pulled out of his pack.

As I listen to Bryan, I think about the places these birds pass through on their way here: of the disintegrating cypress swamps the rufous hummingbirds fly from; of the willow groves that Swainson's thrushes have long sought out in San Francisco's South Bay; of the drowning bayous the ospreys pause in before crossing the Gulf. The birds are all nomads, at home in movement. But what happens when points along their paths begin to disappear? What disorientation will settle upon all of us then? I'd like to think that we can become more and more ourselves through this vicissitude. That through our losses we will be made whole.

＊

Bryan and I tumble down a fern-covered gully. We are on our way to the next stop on the morning's count: point 59. This valley is so large, so verdant, and so seemingly unending that the only useful reference I have is the cartoon *The Land Before Time*. I tell Bryan this and our talk turns to dinosaurs. Back when the dinosaurs lived, the world was a lot warmer than it is today. Sea levels were 550 feet higher. If current sea levels rose 550 feet, they would drown not only the overwhelming majority of the refuges in our coastal flyways but also one-third of the planet's dry land.

"Maybe we secretly love dinosaur movies because they make us think that extinction is somehow reversible," I say.

At that moment Bryan's ankle buckles. He reaches back, his body lengthening, then bouncing against the forest floor. In seconds he is back on his feet. "Damn boots," he says, wiping the pine needles from his pants.

We walk out across a giant Douglas fir bridging two steep slopes, above a brook that connects with the Mack Creek below. The ferns are so thick that it is hard to know exactly how far we could fall. In this case the uncertainty is probably a good thing.

At point 59 Bryan hears two Pacific-slope flycatchers, a hermit warbler, a varied thrush, and a Swainson's thrush, its song spiraling upward. When I ask him how tough it was to learn the birdcalls, he says very. Most have, as you might expect, more than one song. The hermit warbler, for example, has six, a trait that Bryan lovingly describes as "bogus."

Any language learner will tell you that studying a new linguistic system is a humbling experience; with it comes the knowledge not only of different words and grammatical structures but also of worlds and cultures unlike your own. When you learn a language you learn to see your way of life as one of many, your place on this earth as fragile and shifting as any other. Sarah and her team cannot directly ask the birds how they are adapting to climate change. But they can learn a little of the language of the flycatchers and then head to point 400 to listen very carefully for a scrap of that song.

The Andrews is a place not only of long-term scientific inquiry but of deep reverence for the natural world. Each of the scientists working here has learned to see things from another perspective—through, for instance, the eyes of the rufous hummingbird or the groping roots of the Douglas fir. It reminds me of something Robin Wall Kimmerer—who, not coincidentally, served as the Andrews's second writer-in-residence—observes in *Braiding Sweetgrass: Indigenous Wisdom, Scientific Knowledge, and the Teachings of Plants*. "Doing science with awe and humility is a powerful act of reciprocity with the more-than-human world. . . . Love of data . . . the wonder of a p-value . . . are just ways we have of crossing the species boundary, of slipping off our human skin and wearing feathers or foliage, trying to know others as fully as we can."

Bryan has spent the past three summers slogging through this rough mountainous terrain in order to listen to the songs of his fellow migrants high overhead. It is a labor-intensive task, one that demands he pay attention in a world where we do not tend to "pay" anything without receiving something in exchange. But just as paying attention to another person fosters intimacy and makes us feel less alone, perhaps scientific observation allows us to enter into a similar relationship across species. By listening, by returning to the grove time and again, by tuning our ears to the sounds of beings unlike ourselves, we begin to reenter what Thomas Berry, the Catholic eco-theologian, calls "the great conversation" between humans and other forms of life. This too can have a grounding effect, can help stave off a different, larger, and more gaping loneliness. If anything is sacred, it is this, I think. And by *this* I mean all of it: the salmonberries beginning to ripen in the bramble; the scratchy, scolding caw of the Steller's jay that will nibble there; the long, straight trunks of the Pacific red cedars that rise into the sky's blue cathedral. The web of life that too often capitalism seems dead set on dismantling.

From up in the forest canopy I catch a filament of flutelike song. "Is that the Swainson's thrush?" I ask. Bryan nods.

The first step in learning to think of that olive-colored drifter as one of my own, as a member of my scattered tribe, as part of the constellation of people and places I most fiercely defend, is knowing the sound of its voice. The chirps and squeaks it makes to warn of outsiders in the grove, and those it casts into the air to draw another thrush in. Its song of protection, and the melody it sings to reproduce. Like so many of the other beings in the Andrews, it too passed through the marshes of coastal California on its journey north, compelled by a desire to write its name in the sky.

By the time we reach our final stop—point 40—at 10:28 a.m., Bryan has tied his beard up in a knot and the birds are growing quiet. He listens and notes four different species. I sit on a gigantic stump, the body of the ancient tree harvested nearly a hundred years prior. Bryan offers to take my photo, and as I hand down my camera a northern spotted owl swoops through the clearing and lands on a branch not twenty feet away.

The spotted owl's appearance pins me in place. Its eyes are as big and round as shooter marbles, as impenetrable as obsidian. For a full minute it does not blink. The intelligence flickering beneath its downy feathers is somehow wholly different from my own. It is an intelligence that belongs to the old growth in a way that I do not. Unlike all the migratory creatures just passing through the Andrews, the spotted owl will mate and breed and die in place, here. Only here. This forest is the one home this bird will ever know.

I had hoped to see a spotted owl when I accepted this position and have never admitted it, fearing disappointment. Today there are about two thousand pairs of spotted owls left in the Pacific Northwest, four of which are known to mate in the Andrews. During the "Forest Wars" of the late eighties and early nineties, a battle raged over the future of the region's old growth: the lumber industry predictably calling for harvesting and job creation, environmental groups for preservation. Back then it was common to see a "Save a lumberjack, eat a spotted owl" bumper sticker bumping along the forest service roads that run through the Cascades. The recession had put many out of work, and it was feared that the few remaining timber jobs would disappear if the owl was listed as endangered.

But members of the H. J. Andrews spotted owl research team

had been stalking the understory in search of these elusive creatures for over four decades. In the middle of the Forest Wars, they proved that these shy birds reproduce only in old growth, and that the felling of the Northwest's ancient forests was precipitating their decline. A judge used this small fact to create an injunction against harvesting nearly twenty-five million acres of spotted owl territory back in 1990. Today these large swaths of old-growth forest are part of Oregon's $16 billion outdoor recreation industry, which pumps revenue into rural communities that once depended upon timber.

A second bird swoops in and looks right at me. I scribble a string of nonsense in my notebook, a desperate attempt to translate my excitement at experiencing what feels like a moment of interspecies communication. Never before have I stared into the eyes of something so wild for so long. Soon I give up. Settle in to my awe and my ignorance. Transcription will come later, I tell myself, gazing at the birds' brilliant polka-dotted plumage instead. One owl flies toward the other, wings cleaving the air. It lands on the same branch and waits, shoulder to shoulder with its mate, for what I do not know. I watch them for nearly an hour. I want to be alone with these beings near extinction, whose lives are doubly meaningful by our perverse design.

✤

The next afternoon, back at headquarters, I check in with Steve Ackers, the current lead researcher on the spotted owl study. Steve is wearing an all-gray outfit: gray Carhartt baseball cap, gray KEEN clogs, gray T-shirt, and gray quick-dry pants. The day I was invited to be the writer-in-residence at the Andrews, I was teaching an essay in which Steve appeared. At the time the connection seemed serendipitous. But as I talk with him on my last afternoon—recalling my own spotted owl encounter—coincidence is the furthest thing from my mind. I feel like part of the network of beings that make this place what it is and are shaped by it in turn.

"One owl looked at me, without blinking, for over a minute straight," I tell Steve eagerly. Its gaze had felt like such a tremendous gift.

"Do you want to know why they do that?" he asks. "Why they sometimes allow you to stare?"

"Yes," I whisper.

"We've been in contact with the owls for over a decade," he says. "Usually we bring mice."

"You mean when I reached into my bag, it didn't spook because—"

"It thought you were pulling out a box of bait mice. They've become completely habituated to humans. It's the one part of the study I don't like."

I look around Steve's office, which is plastered with photographs of spotted owls, wings outstretched and claws cranked open. Mice had been pulled from Steve's backpack, in part, for this explicit purpose: to capture an image of an enigmatic creature. An image that he would share widely to help keep the Andrews safe. He and the spotted owls have formed a bond, a kind of kinship founded on the reciprocity that comes with paying attention, forged in the fire of mutual exchange. *I'll give you a mouse if you let me take your photograph.* For fourteen years they have been, consciously or not, pursuing the preservation of these groves together. Preserving the rocks that lie under the surface of Lookout Creek, moving downstream with each big storm, remaking the river from beneath. Preserving the high spine of columbines for the hummingbirds that feast there in May. Preserving the pistol-butted hemlock and the tangles of witch's hair. Preserving the branches from which the Swainson's thrushes cast their hopeful song. The loss of its wildness was a price the owl was forced to pay in order to save so much else.

In fact, the logging ban that protected the spotted owl's breeding habitat is—if you look at the results it produced as opposed to the intention behind the action; if you consider, for instance, the varied geography of the places the ban safeguarded and not just the specific species it was designed to save—an example of what biologists have begun to call "conserving the stage." Instead of focusing on the longevity of a single species or place, these scientists suggest setting aside areas rich in geophysical variation. To maintain our planet's diversity of life we need to pay special attention to topography, they say. That's because, as the earth warms, species are on the move, many relocating up in elevation or poleward at a respective rate of forty vertical feet and eleven miles every decade. These species are attempting, in part, to track their thermal niche, the temperature in which they are most likely to survive and thrive. Here in the Pacific Northwest, Douglas fir–dominated old-growth stands can occur anywhere between sea level and roughly six thousand feet. Setting large parcels aside inadvertently meant safeguarding areas across geologically diverse landscapes. And it meant presenting the flora and fauna that live here—at least the mobile kinds—with an opportunity to respond to rising temperatures by rising with them, up the mountainside.

When Steve asks me if I know where I was when I spotted the spotted owls, I say yes. It was 10:28 a.m., and I was standing at point 40 on the long-term bird count survey. Steve doesn't know where point 40 is, so I run over to the bird crew's apartment and ask Bryan if I can borrow his map.

As I walk past the Rainbow Building, my mind turns back to migratory birds and the marsh. Classifying landscape resiliency as a function of topographical variety goes a long way toward explaining just how vulnerable tidal marshes—and the thousands of species dependent upon them—are. After all, most sit within three feet of the highest high tide. This also explains why roughly half of the over fourteen hundred plants and animals considered at risk of extinction in the United States pass through wetlands at some point in their lives. The question becomes, not how we can save the osprey or the rufous, but how we can preserve the various places where they pause during their transcontinental treks, many of which are, unlike the Andrews, painfully flat. The answer is as complex as it is simple. If we want the creatures that make their lives in tidal marshes to have the chance to rise up along with rising temperatures and rising seas, we need to relocate the human communities that restrict the possibility of the marshes' upland migration.

Back in Steve's office I unfold the map and tap my finger over the single black dot of point 40. For a few seconds he says nothing. Finally he asks, "Are you sure they were spotted owls and not barred?"

"I think so," I say, pulling out my camera to show him the proof.

He recognizes the birds immediately. "That's them. The male hatched in 2006. The female is older, almost fourteen now," he says. "We haven't seen them since March. Now we know why. They've moved."

For the first time in ten years, the owls had crossed Mack Creek, moving higher up the mountainside that has long been their home. Only a few months later would I realize what this might mean. I was having a discussion with a literature scholar about Alexander von Humboldt, the Prussian explorer and naturalist known for his observation that as altitude changes so too does the climate. For every two hundred feet up the Andes he climbed, the temperature dropped by a single degree. And as the temperature changed, so did the species.

Later that day, I open a photo I took of the bird count map and compare it with the one the owl researchers use, counting the topographic lines between point 40 and the owls' old breeding tree. Four slender bands separate the two. That means that point 40 is eight hundred feet higher than the owls' former home and that it is also likely a degree or two cooler up there. Perhaps the owls moved because they were trying to track their thermal niche. Perhaps the temperature had slowly begun to rise in the Andrews, causing this bird, one of the most stationary and territorial in all the forest, to migrate uphill.

I write Steve an e-mail, floating my idea. His response is fast and professional: "I think your thermal niche hypothesis has a lot of merit, but it would be tricky to tease out the effects of stressors from a shift in elevation in response to climate change without better data."

At first I am disappointed. Then I open up the images of the owls I had snapped. I look at them looking at me from their perch and I sense that their fate and ours—and the hummingbird's and the osprey's and the columbine's and the cordgrass's and the scarlet gilia's—are all tied up in one another in ways that I can't quite explain. Though I have tried. Tried to make something durable out of language that flickers like the wing of a rufous in flight. Tried to make myself mindful, at least, of the role topography plays in determining a creature's chances for survival.

Often I have wished I were back at the Andrews, though not in that grove staring at those spotted owls, and not at the top of Carpenter Mountain. I have wished myself back into that first light when I learned to hold still and listen to my fellow traveler's flutelike song rising from its branch.

On Restoration

Richard Santos: Alviso, California

I WAS BORN IN ALVISO. MY DAD CAME HERE, LIKE MANY other Portuguese at the time, to work in the canneries and to turn fruit into wine. Eventually he got some underwater land, you know, in the marsh. He had a hard time selling it, so he decided to turn it into a dump. And that became our family business. He had regular contracts, took the waste for most of the town.

There was one thing all of us who lived in Alviso had in common: we were poor. Though we didn't really know it. We all worked in the orchards, and when the days were done we fished off the levees. It was a Huckleberry Finn kind of life.

When I was a kid we'd go down to the salt ponds with Tommy Lane. We'd tie a rope with a stick on the end to the bumper of Tommy's truck and call it waterskiing. That Tommy was a champion swimmer. He probably has about one hundred trophies back in his house. When I was in high school, in 1959, the coach from the Santa Cruz swim club came down. He said to Tommy, "Son, I'm going to take you all the way to the Olympics." Tommy, being a bumpkin from Alviso with an IQ of about seven, said, "No thanks. I'm gonna marry Beverly and work at the power plant across the street." And that's exactly what he did, worked there for forty years.

Everything I know about fishing I learned from Tommy. Back in 1979 I caught the biggest sturgeon in the history of Alviso. Took an hour and ten minutes to reel it in. Seven-foot, 900-pound sturgeon. When I finally threw the fish in the boat it busted up the place. So I threw it back. Years later I caught a 207-pound sturgeon. We took that fish to the local Chinese restaurant. There were about thirty-seven of us. We had them sauté it with sweet-and-sour and onions. What a feast! We gave them the head and they cut open the gills and big chunks of meat fell out.

Back then it was mostly salt ponds and orchards. There was no Highway 237, only Route 9. People thought this was the outback. Judges, congressmen, governors, they all came out here to Vahl's Restaurant—which is still operational, by the way, her nephews run the place. Vahl's, and all of Alviso really, was so secluded, so far from everything else, that there was little in the way of law. People drank, they gambled, they whored. The trains all ran through town and the hobos would jump off, and my father—he eventually became the police chief—he would greet them right there and ask, "Are you working?" If you said no, he put you in a car, drove you to the outskirts of town, dropped you off, and told you never to come back. Back then this was all migrant camps—Portuguese, Armenians, Spaniards, blacks, okies. Everyone was generous. It's often folks with the least that share the most. I would sit by the fire at the camps and listen to them singing and they would feed me tortillas.

And now everyone is trying to get a piece of Alviso, and I won't let them. I don't want high-density housing and tech campuses. I want my grandkids to be able to see the mountains and to run along the levees. Sure we've flooded in the past, but I have a sixth sense about flooding and I don't think we're going to flood again, mainly because that wetlands restoration project is providing a lot of additional protection that we never had before. And it's helping to bring back all the different kinds of fish we had out here when I was growing up.

The thing is, we the people of Alviso own this city, and nobody with big money owns it. A couple years ago a developer approached me and said, "Mr. Santos, you have four acres. I will give you one million for each of them." They wanted to put a couple hundred houses in there. I said, "What am I going to do with that four million bucks? I'm going to move to Beverly Hills. Get two German shepherds. Sip a beer and eat a sandwich. But who am I going to talk to? I can't go back home anymore because I sold out. And I can only buy so many sandwiches and beer, so how the hell do I find happiness?" He thought there was something wrong with me. Well, I thought there was something wrong with him.

There are a lot of people in San Jose, with money, who would like to see me disappear. Only because they would get their way with development. I fight so hard because I'm trying to preserve the history and the characteristics of Alviso. The stories we tell about this place are powerful. More powerful than money. More powerful than all the different fantastical futures the tech industry might offer. When you take a cubicle and you put fourteen mice in there, they destroy each other. But if you put four in, they survive. Now that cubicle is Alviso. Alviso can exist. And you can bring business in. But you can't make it too tight. The City of San José doesn't care. They just want a slam dunk.

A couple of years ago the Alviso advisory board approved a development of office buildings on the southern end of town for Cisco. Now years go past and the dot-com bubble crashes and Cisco can't afford to build anymore. So they sell the land to another developer that wants to put in a big manufacturing and trucking distribution center. But we don't want that traffic rumbling through our streets at all hours of the night, right next to our school. That's not the kind of development we approved. Well, the City of San José told us that we had approved a project, and they gave the developer the go-ahead. They just steamroller us as they have so often in the past.

But I'll tell you what, the wetlands restoration project out there in the bay, they're keeping huge chunks of land out of developers' hands, and that keeps the area around Alviso open space. And when that area is open space, it can absorb the floodwaters that would otherwise run right through this town. They wanted to build a stadium out in those wetlands, but now that land is part of the project, so it is illegal to develop them. For me, that's the real slam dunk.

When I go out into the area around the Don Edwards Center I see plover, egrets, burrowing owls, harvest mice. I hear the grass shrimp are back too, though I don't get around to fishing much these days. They all seem to be more plentiful than they were a decade ago. And I tell you, lately there are a whole heck of a lot more jackrabbits running around than I've seen in a long time.

Those wetlands are going to save the community that I've fought so long to protect. I was skeptical at first, sure. I thought it was just a bunch of environmentalists mucking around in the weeds, but now I see it's the best chance we've got to keep Alviso *Alviso*. To save the community and the ducks and geese and sturgeon and all those other animals that have made this a wonderful place for me to live out all of my seventy-some-odd years.

Looking Backward and Forward in Time

San Francisco Bay, California

"THE SCALE OF WHAT WE ARE PROPOSING TO DO OUT HERE scares people," says John Bourgeois, the executive project manager of the South Bay Salt Pond Restoration Project, the single largest wetlands rehabilitation effort this side of the Mississippi River. It's 2017 and John and I are standing in the Don Edwards Environmental Education Center in a glass-walled widow's walk. To the west is the tiny working-class town of Alviso. Just beyond it Silicon Valley spreads, its hunger turning earth to glass. Together we look out over the trailers and the former canneries, over the shimmering tech campuses—Dell, Google, and Microsoft—and the slow slope of the nearest landfill. In a way, San Francisco's South Bay is a microcosm of the American coast, with everyone from young Facebook execs to sanitation workers holding a slice of the sinking pie.

Directly in front of us, the former salt ponds from which the project takes its name dominate the landscape. "At the northeasternmost corner is pond A1. *A* is for the Alviso Complex and the number one means that was the intake pond. That's where Leslie [Salt Works], and later Cargill, allowed bay water to enter the system," John says. Over time, as the water evaporated, the salt levels rose. "Then they pumped that water into pond A2, where it evaporated some more, and from A2 into A3, A4, A5, all the way down to pond A22. It took five years for a molecule of salt that entered the system to get harvested."

Today some of the ponds are open water, while others are covered in a green patchwork of pickleweed and bulrushes. Most haven't been used to make salt for well over a decade. Weeks of heavy rain have left the earthen embankments that divide one pond from the next so unstable they can't be trusted under the weight of John's Prius. "We're going to have to walk instead," he says. "But first I wanted to give you a kind of bird's-eye view of the project, or at least this section of it." John has slate-blue eyes and extraordinarily long eyelashes, and, like every other male professional I have met in the Bay Area, wears trail-runners, jeans, and a button-down shirt, which he untucks ten minutes into our meeting.

John was born, raised, and educated in Lafayette, Louisiana, a place where wetlands define daily life, even as they disintegrate. "I used to have to drive two hours to get to the boat launch, then motor the boat an hour and a half just to arrive at my old project site," he says. "When I came to San Francisco Bay I was like, 'Where *are* all the wetlands?' I couldn't believe I was looking at the biggest estuary on the West Coast. I could toss a stone across the largest extant tidal marsh. That's when I realized just how profoundly devastated and fractured the San Francisco estuary was, just how much of it had been lost to development."

Reach back two hundred years and the scene that John and I look out over would be wholly different. The levees gone, the open water gone, the mauve tint of microbial activity in the few still-functioning salt ponds gone. Imagine instead a sea of grasses in all shades of green, from rich emerald to the misty gray of tule and fog, from honeyed lime into the blown-out colors of dried papyrus. Willow trees would be flourishing along freshwater creeks, a ribbon of spotted sandpipers in from South America fluttering over the mudflats. If you can imagine this, then you can at least partially imagine what John is trying to bring back.

While the indigenous people of California long used the southern spur of the bay to produce salt, it wasn't until the 1850s, when nonnative, family-run operations began to sprout up, that the tidelands were dramatically transformed. Salt was a hot commodity in the Wild West, vital for both the preservation of food and the mining process that drove hundreds of thousands of prospectors to the Sierra Nevada. So fierce was the demand that local "salt makers" began to alter the bay's low-lying areas in an attempt to speed up the procedure. They paid laborers to heap dirt along the bayside edges of the mudflats, restricting tidal flow and accelerating evaporation rates. Levee after levee went in, and the rhythmic rise and fall of the bay water through the mudflats and marshes ceased. Over time, Leslie Salt Works took control of each of these relatively small-scale outfits. The company acquired one salt farm after another until it owned over forty-four thousand acres, an area equal in size to three Manhattans, on the east and west flanks of the South Bay.

In 2003 the state of California purchased many of these salt ponds from Cargill, Leslie's successor, paving the way for the most innovative and forward-looking wetlands restoration effort in the country. Since then the project area has grown in size, and today it encompasses about fifteen thousand of the forty thousand acres of former tidelands that the Bay Area is attempting to rehabilitate. But climate change has significantly complicated the endeavor. When wetlands restoration began in California the goal was nostalgic: to simply return things to how they used to be. Now it also involves figuring out how to transform one of the flattest and most vulnerable landscapes on the planet into something more robust, into a place that just might make it into the next century.

Its scale separates the South Bay from every other coastal resiliency project I have covered. In Rhode Island, for example, the United States Fish and Wildlife Service and the Nature Conservancy recently sprayed sand over a decomposing, anvil-shaped tidal marsh at Sachuest Point. Their intent was to raise the marsh to help it keep pace with rising seas—but never mind the road running along its upland edge, impeding marsh migration, and never mind the nearby rivers, dammed by centuries of development, whose sediment no longer replenishes the eroding land. I walked away from that site visit disappointed. The marsh looked like a moonscape, alienated from the geophysical processes that might help it keep up with the rise in the long term. The size of the parcel (eleven acres) and the adjacent land-use practice (tourism and waste disposal) determined what would remain beyond the project's scope.

After my morning at Sachuest Point, which followed directly on the heels of my month in the Andrews, I started to wonder whether anyone was considering how we might rehabilitate, nurture, and maintain vast swaths of our tidal wetlands not just for ten or twenty years but for a hundred. Whether anyone was attempting to reconnect our imperiled coastal landscapes to the rivers and forests that had long supported them. Whether there was a single enigmatic species, perhaps, that might demand such wide-ranging interventions. Or whether a different approach to conservation was brewing, given our increasingly uncertain future.

First I found an indigenous community building "pocket estuaries" along Puget Sound, intended both as rest stations for the migrating Chinook salmon and to slow surges of storm water into the Swinomish's nearby village. It was a start, but a relatively localized one. I circled back to dozens of coastal communities preparing to pull up their roots and move in: the Hoh and Quinault of the Olympic Peninsula; the towns of Newtok and Shishmaref, Alaska; parts of Sayreville and Lawrence Township in New Jersey. But I was looking for something larger. A project whose size alone would draw controversy. And then I found Measure AA, which passed in June 2016 with nearly 70 percent of Bay Area voters agreeing to pay a twelve-dollar-a-year parcel tax to fund wetlands restoration. Soon I was calling reporters, politicians, activists, and scientists. When I ended each conversation with my standard request—"Please connect me to two other people with whom you think I should speak"—more often than not I was told to contact John Bourgeois.

So now I am standing beside John, looking out over a climate change adaptation strategy unlike any I've ever encountered. At fifteen thousand acres, the South Bay Salt Pond Restoration Project is one of the largest coastal wetlands rehabilitation efforts in the country, second only to the work taking place in the Everglades. The project's staggering size enables John and his team to experiment with landscape-scale interventions that have never before been attempted. At least not in coastal wetlands. And not with the express purpose of attempting to bring back what has been lost while readying it for a future we don't really understand.

✳

This is how Hoover built the dam: It begins with a general rule that the land east of the Mississippi gets enough rainfall to support agriculture, while land to the west does not. First there was a decade of drought, then a decade when settlers on the frontier realized that most of the homesteads with reliable water had already been snatched up. But that didn't stop them from continuing to cross the country. Over the fifty years surrounding the turn of the century, the population of Los Angeles increased a thousandfold, the population of Las Vegas by roughly three times that. In 1902 the Reclamation Act passed Congress, providing federal funding for irrigation projects that would turn desert into farmland. The first two decades of the program were mired in failure, but this meant only that the size of the next project expanded, like a snowball on a steep slope. In 1922, Secretary of Commerce Herbert Hoover brokered a deal to divvy up the flow of the Colorado River into unequal shares, one for each of the seven states it flowed through. Less than a decade later, the scrappy Six Companies consortium won a bid to construct what was then called Boulder Dam for $48,890,995.50. Tunnels were blasted through the bedrock, the Colorado River temporarily rerouted through canyon walls. In about as much time as it takes to earn an undergraduate degree, the largest concrete structure in the world was complete. And with it began the large-scale transformation of one of the West's wildest rivers into a fourteen-hundred-mile-long canal. Today the Colorado provides water to 36 million people and power to 1.3 million, irrigates 6 million acres of farmland, and generates $26 billion in revenue yearly thanks to the recreational activities enjoyed along its banks. There are eight major dams "controlling" its flow. They have been so successful that since the early 1960s the river has rarely reached its natural outlet in the Gulf of California. Instead it sputters to a stop somewhere in northern Mexico. Over the last decade, the Colorado's groundwater has started to disappear and the water levels in many of its reservoirs have dropped by hundreds of feet. Meanwhile the population continues to balloon, the drought deepens, and the temperature rises.

✻

Unlike the East Coast of the United States, where the deglaciation process has left much of the shore sloped low, the West Coast rests along a seismic fault line. For over a thousand miles, hard-packed clay and serpentine are stacked vertiginously over the sea, in what we call the Coastal Range. These mossy mountains extend from Alaska all the way to Mexico, and between the Columbia River and Baja there is but one break in their rampart.

Beneath the Golden Gate Bridge, the Pacific Ocean rushes inland, filling three kidney-shaped bays. There the salt water mixes with the Sacramento–San Joaquin River delta and the freshwater that drained from the western side of the Sierra Nevada. For millennia prior to the arrival of Spanish missionaries, hundreds of thousands of acres of wetlands ringed these bays in a rich ocean of grasses. Over fifty distinct tribal societies inhabited the lands along this sheltered section of the coast's otherwise rocky edge. Here they harvested oysters and clams; here they caught sturgeon and hunted for waterfowl.

All it took was two centuries to transform what had once belonged to no one, and so to everyone, into a commodity—to be bought and sold and farmed and stockpiled; to be made to produce more salt than any other place on the planet; to be turned into low-income housing and glittering tech campuses, into tens of dozens of wastewater treatment plants and planned communities with picket fences and golf courses and fake stone fountains. As the cultures of the Bay Area's indigenous people were almost completely destroyed—first thanks to the missions and disease imported by Europeans, and then through the forced labor that came with Spanish, Mexican, and American occupation—the idea of who had access to the region's wetlands mutated.

Only three years after American soldiers seized the Mexican province of California, the Swamp Land Act of 1850 passed Congress, granting new states the right to sell flood-prone land to individuals as long as the water could be drained from it. The legislation sparked the nationwide annihilation of wetlands from the Florida Everglades to the San Francisco tidelands. The Swamp Land Act was a classic form of the get-rich-quick scheme that defined the colonial project: steal indigenous lands, auction them off to the highest bidder, and then enforce property taxes, guaranteeing a long-term source of funds. It would transfer the promise of future financial security away from the country's first inhabitants. In just two years' time, nearly 790,000 acres of California's wetlands were shifted into the hands of fewer than two hundred private owners, who, having paid the bargain rate of $1.25 per acre with no money down, proceeded to dam, dike, drain, and fill the largest estuary on the Pacific coast.

And so the frenzy began. Swamplands were leveed off, irrigated, and turned into fields for stone fruit and avocado trees. Swamplands dug out and made into privately owned oyster beds. Swamplands transformed into commercial strips, including sections of the Embarcadero, which defines downtown San Francisco today. Swamplands turned into the largest salt production complex in the world. Swamplands filled in to support residential communities in Oakland, Richmond, Union City, and Fremont, several of which would provide relatively affordable housing well into this century.

Over the past two hundred years, 90 percent of the region's former wetlands have been twisted into shapes so unlike their old selves we often fail to recognize them. Crissy Field, the old army dump turned aerodrome turned beloved Presidio parkland? Formerly Yelamu-occupied salt marsh. East Palo Alto, one of half a dozen neighborhoods that thrummed at the hot center of the black power movement? Plunked down in a sea of bull grass. Alviso, where Jack London met the dog that likely inspired *The Call of the Wild*? Former mudflats.

I drive through Alviso on my way to meet John at the Don Edwards Center. Immediately it reminds me of Oakwood Beach: the working-class housing stock, the salty-sweet air that sweeps through the streets, the feeling of being somehow removed from the bustle of big-city life. Most importantly, like Oakwood, Alviso juts out into the surrounding marsh like a swollen thumb.

"They might make good candidates for relocation," I say later, as John and I walk a levee that runs along Alviso's northernmost edge. Much of the town has subsided significantly over the past century, thanks to the groundwater extraction necessary to support water-hungry crops like peaches and almonds. Together we peer into the hollow second story of the former Bayside Canning factory, with its radiant archways and hand-painted murals of women with long black braids carrying fruit to the production line. Most of the windows are gone and so are many of the walls. My gaze rests on the roof gable of a residential home just beyond the ruin. If I look left, I am standing five feet above the bay; if I look right, I can peer into this attic. That is how much the land has sunk since Alviso's founding.

John does not acknowledge my relocation comment. It is not the first time, nor the last, that one of my inquiries into potential retreat in the Bay Area is either shot down or ignored. To entertain the idea is to acknowledge that even the largest wetlands restoration project on the West Coast might have a limited life span.

"Do they flood?" I ask instead.

"The entire town is sixteen feet *below* sea level," he says.

"Yeah, they flood. From the river side and from the marsh. But they haven't in a while because we don't tend to get big storms. Plus all the old Leslie-built nonengineered levees, like this one, that surround the salt ponds, they provide informal protection."

To return tidal flow to the area, John has to remove the earthen embankments that separate the ponds from the bay. But before he does that he has to build a four-mile-long replacement levee running all the way from Baylands Park, around Alviso, and over to the Newby Island landfill. "The Army Corps of Engineers came up with a recommended levee height of twelve and a half feet, but their estimate didn't take sea level rise into account." John pauses and gives me a look meant to communicate the sheer absurdity of leaving climate change out of any analysis used to design future flood protection. "My employer, the Coastal Conservancy, and the Santa Clara Valley Water District decided to band together and pay the additional $16 million it would cost to make the levees sea level rise ready. That means going up a couple extra feet and making them considerably wider as well."

"Wider?" I say. The word stops me in my tracks. "Like how much wider?"

"Some are going to be built at a thirty-to-one slope, others are going to be one hundred to one." A ten-foot-high levee would, in the first case, be three hundred feet wide, and, in the second, a thousand. In the most basic sense, additional levee width increases topographical diversity in a landscape that currently has very little. "We need the width to build in transition-zone habitat," John says. "Come down here during king tide and you'll find hawks cruising the levees, picking off endangered species. In the future it's only going to get worse, the breeding and foraging habitats of these animals squeezed between the sea and the surrounding communities."

The more John speaks about these nearly horizontal levees, the more excited he gets. The current goal of the South Bay Salt Pond Restoration Project is to restore as much wetlands habitat as possible before 2030, when sea level rise is expected to accelerate significantly, and to line, wherever possible, the upland edge with horizontal levees. John hopes that they will give the tidal marsh the chance, at least in the short term, to migrate up and in. It is one of the first times in nearly half a decade of writing about wetlands communities that I have encountered someone who is physically doing something that might help the nonhuman species dependent upon these low-lying and inherently vulnerable landscapes to adapt to rising tides.

"I even have a dirt broker," John continues. "Silicon Valley is booming again, people are building like crazy. That's a lot of dirt, and a lot of that dirt ends up in landfills. But now, instead of paying the landfill to dispose of it, construction teams are giving it to Pacific State [a local aggregate supplier], which tracks the material, tests it, and brings it to us. We're getting two million cubic yards of fill—clean, confirmed, compacted—placed exactly where I want it, for free."

As I listen to John speak, I realize that I have been thinking about Alviso from a distinctly human point of view. I have been imagining myself standing behind the levee, waiting for the water to rush in. John, on the other hand, looks at Alviso from the perspective of its wetlands. He is concerned less with human communities and our ability to adapt—we can and we will, he says—and more with how to keep the 50 percent of endangered and threatened species endemic to North America's wetlands from going extinct. His priorities are to protect the salt marsh harvest mouse and the California Ridgway's rail living among the pickleweed and bull grass. And if that also buys Alviso more time, all the better.

"If we don't have tidal wetlands on the West Coast then we aren't likely to have Ridgway's rails either. It's as simple as that," he says.

"Is there someplace I can go to see a horizontal levee?"

"You have to make it out to Oro Loma. It's a test case for the concept. They're also running treated wastewater through the levee, because you-know-what flows downhill and all our sewage treatment plants are at the bay edge. We're thinking maybe we can even use the human by-products to provide nutrients to the plants that grow on the ecotone."

I write, "ORO LOMA!" in my notebook. Next to "Drawbridge!"—a ghost town still sitting in the middle of the salt marsh near the sewage treatment plant. And "Measure AA!"—the parcel tax that will fund widespread wetlands rehabilitation and resiliency via projects like John's horizontal levees. Over the next twenty years he expects the tax to raise nearly $500 million.

"They spent $1.3 billion building the new 49ers stadium. If I had *that* amount of funding I could restore all of San Francisco's wetlands and build horizontal levees at the upland edge of each one," John says, shaking his head. "I guess we value football, or making money on our love of football, more than our..." He trails off.

❋

This is how Neil Armstrong prepared to walk on the moon: First he lay down inside a three-ton metal ball at the end of a fifty-foot arm. A technician flipped a switch and the steel centrifuge was whipped around a circular course at eighty-eight miles per hour to simulate the g-forces that would be exerted on him when he exited Earth's atmosphere. Or so they thought; no one, of course, had ever been in Armstrong's position before. The gondola allowed the test subject to be oriented in different positions: sometimes Armstrong rode with his "eyeballs [facing] out," sometimes "eyeballs in." Either way, if he relaxed his muscles his vision would narrow like a set of blinders and he'd black out. Then he practiced piloting a replica of the lunar landing craft. A jet engine thrust the pod up, up, up, 250 feet above California, above the Mojave Desert and Edwards Air Force Base and the world's largest compass rose, painted on the dry bed of an ancient lake. Then the engine pulled back to a near hover, supporting all but a fifth of the machine's weight. Armstrong throttled the rocket boosters as the module fell very slowly toward the earth. This is what it would be like to land on the moon. Or so they thought. Finally his body was loaded into six slings and suspended perpendicular to the ground. This made Armstrong feel almost weightless. He walked and jumped, ran and fell all along the walls of the Lunar Landing Research Facility. Sometimes he did it wearing a heavy backpack, sometimes not. This is what it would be like to walk on the moon. Or so they thought.

✷

The night before I first visit the South Bay Salt Pond Restoration Project, it is drizzling. As I walk along the Embarcadero, past Hog Island Oyster Company and the San Francisco Ferry Building, past Starbucks and the Slanted Door, I think about how for the overwhelming majority of the last fourteen thousand years—the length of time that humans have inhabited the area—this land didn't have a price. I pass a cocktail bar called Hard Water, "a premier San Francisco dining experience." I pass a coworking space and an architecture firm. Eventually I slip into the lobby of the Exploratorium, San Francisco's sleek take on the science museum. Mary Ellen Hannibal, a friend of a friend and a fellow science writer, told me to start my investigation here, with a visit to the museum's Fisher Bay Observatory, covertly designed— she said—to introduce guests to sea level rise and its potential impact on the Bay Area. It is Monday and the Exploratorium is technically closed, but thanks to Mary Ellen I have a private tour.

When curator Susan Schwartzenberg appears from behind a pane of smoky glass I know right away that I like everything about her. Her gray-rimmed round spectacles, her quiet enthusiasm, the way our conversation almost immediately and somewhat inexplicably touches on the critic John Berger's recent death. We both reveal that his book *Ways of Seeing* helped us to understand that the act of observation is always, at its heart, reciprocal. When you see, you are also being seen. And being seen means entering into a relationship with whoever or whatever is looking at you. I tell her about Steve Ackers and the spotted owls in Oregon's Central Cascades, who remind me that paying attention across species is a form of prayer. In return she tells me the following:

This is how Frank Oppenheimer founded the Exploratorium: First Frank was brought to trial by the House Un-American Activities Committee. Then he was forced to resign from his position as a nuclear physicist at the University of Minnesota. Barred from conducting research and blacklisted from holding an academic position, he retreated to rural Colorado to become a cattle rancher. But Frank refused to quit science altogether, so he taught it at a local high school. Instead of using a textbook he filled his classroom with steel cables and cardboard boxes, with plastic cones and floor fans, so his students could construct experiments of their own and practice the skills of scientific observation. Years passed. The senate voted to condemn Joseph McCarthy for his "inexcusable . . . reprehensible . . . vulgar and insulting" conduct, and Frank moved to Fog City to open a new kind of science museum dedicated to "fostering awareness," to seeing and being seen. He raised the funds himself, touring the country with his homespun experiments, hitting up his famous physicist friends and local artists. The doors to the Exploratorium opened without fanfare (Frank didn't have the money to host a big party) a few months after Neil Armstrong placed his feet firmly on the moon.

Together Susan and I walk the length of the building, which is cantilevered out over the bay. The observatory rests at the farthest end. A box of glass surrounded on all sides by water. I gaze into the tidal bellows of the steel-gray bay. Two container ships, their bellies heavy with goods, glide away from the Port of Oakland, riding the outgoing current. They pass a tiny-by-comparison coast guard vessel and a navigational buoy blinking green. A flock of fist-size birds pepper the sky, and a seal pup noses toward the shore. The vastness of the bay, and of all the locations tied to it by trade routes and the transcontinental treks of migratory birds and ocean mammals, is instantly palpable.

"In designing this room, I wanted to encourage people to use the exhibits to become better observers of what's out there," Susan says, turning me around and pointing me toward the opposite view. The skyline of downtown San Francisco—the Transamerica Pyramid, the Salesforce Tower, and 555 California Street (of which Donald Trump owns a 30 percent share)—shimmer with great self-importance. "If you were standing here in 1848, before the gold rush happened, you would be standing twice as far out into the bay. The old shoreline runs along Montgomery Street, and everything east of it was marsh or else open water."

We look down at a table where a map of the preindustrial shoreline is backlit. Susan flips through a box of translucent overlays, gently placing one on top of the map. "You can see by 1853 the city is already beginning to eat into the bay. They started filling in the area around the piers with all different kinds of debris and junk. Mining Telegraph Hill and dumping the dirt from it into the water." She places another layer onto the map and the shape of the shoreline shifts again, expanding outward. "Very quickly, by 1859, they are building the first big seawall at the far end of the piers. You can see it is just a rubbly thing. But it sort of sticks, and by the 1870s we get to a version of the city's coast as it is today."

As I look back and forth between the city and the maps, I am reminded of a video I saw in which salt water laps up and over the Embarcadero during a king tide. The sidewalk is submerged and commuters walking to work appear confused. The Bay isn't supposed to be here, their faces seem to say. But according to these maps, the real mistake lies in their idea of where the shoreline belongs.

Their expressions made me think of the words of Wendell Berry, so simple and forthright, at the opening of *The Unsettling of America: Culture & Agriculture*. Large-hearted and cantankerous, he writes, "Our land passes in and out of our bodies just as our bodies pass in and out of our land. . . . As we and our land are part of one another, so all who are living as neighbors here, human and plant and animal, are part of one another, and so cannot possibly flourish alone. . . . Therefore . . . our culture and our place are images of each other and inseparable from each other, and so neither can be better than the other." What happens when we lose sight of the way our culture mirrors the land? What happens when we lose sight of the land altogether? When we pave over the wetlands that once bore so many fruits, that once brought fifty different tribes together through a common resource, that once offered flood protection during the rare storm that spun into the San Francisco Bay? Berry would say that we suffer. Our culture suffers. All the plants and animals that once called these places home suffer. Doubly so now that the sea is rising. When I think of Wendell Berry and the Embarcadero, open water less than two centuries ago, I can see what ought to be obvious: we shouldn't be standing here in the first place.

Susan sees the gears in my mind rotating and notes, "Even today, if they're digging to build some kind of new foundation in this area, it isn't rare for them to come across a buried ship." We peer into a case filled with a hodgepodge of found objects: pearly teacups, blown blue glass, bits of bones, and empty bottles of French champagne. Each was taken from a boat or a storehouse some builder unearthed.

Susan has one final piece to show me before we leave the Fisher Bay Observatory. We walk together to a car-size topographical map of the Bay Area in the center of the room. Filtered blue light is projected over the miniaturized landmass, indicating the current shape of the shoreline. "This is where we invite folks to think about sea level rise," she says, turning a knob at the base of the exhibit. But she doesn't spin it forward, moving the projection into the future; instead she pulls the projection backward through time. "This is eighteen thousand years ago," she says. Back then much of North America was covered in a massive sheet of ice. Sea levels were about three hundred feet lower, and the Bay Area's telltale kidney-shaped bodies of water all but disappear. Susan slowly turns the knob forward and the projection travels toward the present. There is little variation in the shape of the shore for roughly five thousand years, then all at once it changes, the ocean rapidly covering up a significant area of dry land.

"Wait," I say. "Can you go back?"

Susan obliges and I watch the water pull away. Then she spins the dial again and the ocean throbs inland. I ask her to show me the transition once more, but this time I don't watch the water—I watch the key instead. Just as I suspected, the jump occurs between twelve and thirteen thousand years ago, during the event glaciologists have labeled Meltwater Pulse 1A. The event that Hal

Wanless, back at the University of Miami, points to when arguing that sea levels tend to change abruptly and with great speed. In the time it has taken me to write this book, the predicted rise by century's end has doubled. If sea level rise continues to accelerate at even half this speed we are looking at an increase of well over ten feet in the next eighty years.

Susan moves the map into the present, and the colors of the shoreline shift. Any area that once was bay and has been filled is a deep, conifer-colored green. Roads are pink and airports are orange. As Susan turns up the water levels—adding one foot of rise, then two and three, four and five—big sections of the city begin to disappear beneath the blue light.

"It's totally freaky to see that with just two feet you have lost two airports. Pretty soon all of the approaches to all of the bridges are gone, most of Highway 101 and Route 37, and healthy chunks of the 80." The few current pockets of afford-ability—Alviso, Redwood City, Fremont, Richmond, and East Palo Alto—are all underwater, as are portions of Oakland, Marin County, and downtown San Francisco. What we think of as the coastline is a blur. Looking over the map, I wonder whether the Fisher Bay Observatory is designed to get visitors not just to think about sea level rise but to begin to imagine the unthinkable: unsettling the American shore. Relocating the highways and the bridge bases, the Slanted Door and the Starbucks, the communities of color and the former cannery towns, the homes in Mission Bay, the Facebook campus and the Google campus and all the landfills and the sewage treatment plants. I wonder whether this exhibit is a first step in a much larger trek toward letting go of the static image of the coast that we have spent the last two centuries developing and attempting to defend.

*

This is how Frank's brother Robert built the bomb: It started as an accident. Two German scientists were showering uranium nitrate with lukewarm neutrons when the atom suddenly split in two. The world went to war less than a year later, and what had been a tool for producing previously unimaginable quantities of energy became a potential weapon. J. Robert Oppenheimer's old adviser encouraged him to start researching fast neutron chain reactions. Eight months after the United States formally entered the fray, Robert was named the "coordinator of rapid rupture" for the Manhattan Project. The government dumped $2.2 billion into developing the science and facilities it required. They employed 130,000 men and women at the height of the bomb-building effort. It took 44,900 people to build and run Hanford Works, home of the first large-scale nuclear reactor, in Washington. Dozens dug the sewer ditches at Los Alamos National Laboratory, where Robert was in charge. Others felled trees. Cleaned toilets and desktops and bunks. Delivered mail and put out fires. Worked in the hospital and babysat for scientists. Twenty-four died there. Four from drinking muscatel wine laced with antifreeze. One from drowning. One tractor mishap. Multiple vehicular deaths and one suicide. Two died after accidents during "criticality" testing with a softball-size pit of plutonium later nicknamed the "demon core." The same type of plutonium was dropped on Japan. There it went by the name Fat Man. In Japan our relationship to this plutonium was not experimental. And so Robert did not know whether to celebrate or to mourn when in the days and weeks following detonation the throats of the Japanese began to swell, their hair to fall out, their gums and rectums to bleed, their skin to slough off, their organs—lungs and liver and intestines—to dissolve, and, as some said, fall from their mouths onto the floor.

The afternoon following my visit to the Exploratorium, John and I walk an earthen embankment bordering one of the former salt ponds. Half a dozen comb-shaped islands sit in the middle of the water. Their unique silhouette, decided upon after a couple of years of field testing, allows breeding birds to shelter from the wind. In the center of each island, solar-powered speakers play the mating call of the Caspian tern.

Everywhere I look I see another set of human interventions intended to mimic a healthy tidal landscape. To trick the passing migrant into staying and mating. On the one hand, the tremendous effort that has gone into restoring and readying these wetlands for sea level rise strikes me as both unprecedented and wise; on the other, it all seems like too much somehow, yet another example of first-world exceptionalism and the delusion that we can design our way out of the planetwide geophysical transformation we've set in motion.

"Over the last one hundred and fifty years we've changed the shape of the shoreline in the United States, diking, draining, filling, and developing our tidal wetlands," I say, thinking of the map at the Exploratorium. "Do you ever take a step back and wonder if you aren't doing the same thing—I mean, to make a horizontal levee or one of those islands, you do have to dump dirt into the bay—with a different set of information?"

John laughs and says, "Huh," as if he is surprised. Then he asks me to clarify the question. But before I get the chance he speaks.

"If you want to talk about landscape-scale change, all you have to do is look to Louisiana. The whole southern edge of the state is melting away, gone," he says. "We're trying to restore natural processes so these wetlands might escape that fate. Is it risky? Yeah. But I think doing nothing is even more so. Since the project's start, bird populations have more than doubled. We are making the best decisions we can with the knowledge we have. And we have a lot of knowledge. Is this perfect? No. Would I prefer to move Alviso out of the way and let the marshes migrate inland? Sure. Would that be a better solution? Absolutely. But politically it's not feasible."

"There are places that are retreating," I say.

"Really?" he asks.

I tell him about Oakwood Beach and the Isle de Jean Charles. For a moment his mouth goes slack. The Coast Starlight's twelve silver cars slide through the marsh like an arrow through the sky. It always amazes me how little those living and working in these tidal wetland communities know about all the others addressing similar challenges in different locations.

"The thing is, land in the Bay Area is some of the most valuable per square foot in the country. I don't see us giving that up anytime soon," he finally says. It is a sentiment that will be repeated regularly on this research trip, by nearly everyone I meet, from the former executive director of the San Francisco Bay Area Conservation and Development Commission to the District One supervisor in San Mateo County.

John and I pass a notch in the levee where salt water from the bay enters the ponds. From beneath the bridge's steel grating comes the burble of tidewater in motion, and from out in the middle of the pond comes the Caspian tern's raspy call, its insistent *kowk, kowk, kowk*. I can't tell if this is an actual bird or a recording. Either way, when I look out toward the man-made islands, I see the flutter of dozens of charcoal-tipped wings.

"Do you ever fear that you're playing god?" I say, trying out a different version of the same question. "I can see your work in Google Earth."

"We have dozens of different stakeholders and every time we want to move forward we have to get them all to sign off—the Santa Clara Valley Water District and the Sierra Club, the Silicon Valley Leadership Group and the California Waterfowl Association, the City of San José Environmental Services Department and the Bay Institute, residents of Alviso and Hayward—and that alone, that coming together to meet a great challenge, makes me confident that what we are doing is worthwhile. Creating consensus across organizations is, I think, a form of resiliency in and of itself. When we need to come together to solve future challenges, the web of connectivity will already be in place." As compelling as John's words are, they still don't answer my question. I am left uneasy and unsatisfied.

That night I dream that I am a refugee on an island. There are many others like me, displaced and dispossessed, all sleeping on the floor of an empty teak mansion by the sea. One day a massive storm starts; blue black, it spins a set of twisters up from the water's frothing. A flock of birds, hundreds of thousands of them, papers the purple sky. When I see them fleeing, I know—as I have known little else in this life—that having somewhere to retreat to is the key to survival. I also know that I too need to leave. But where can I go? I grab my mother and father by the hand and we run through the building and out the other side. We run down the long arm of the dock and jump into the adjacent marsh. It is high tide and most of the grasses are covered in water. An extraordinarily large wave rolls in and the horizon disappears. Now I have a pair of bolt cutters in one hand, pruning shears in the other. When the storm is on top of us we take a big gulp of air and dive under. Then I wake up.

＊

This is how Harriet Tubman led hundreds of slaves to freedom:
First she freed herself, fleeing through the marshes and swamps
called Blackwater, where she had once collected muskrats from her
masters' traps. Then she changed her name from Araminta Ross to
Harriet Tubman. After a few years in the north she turned around.
Harriet always crossed the Mason-Dixon Line in winter, when the
nights stretched taut across the dormant land, keeping those who
owned homes within them. When she arrived at a plantation she
often disguised herself as a slave and communicated in code, singing, "Swing low, sweet chariot, coming for to carry me home." The
chariot was the Underground Railroad, and home was freedom. If
you heard this song, the time had come to run. She always sang on
Saturday, since Sunday was a day of rest and newspapers wouldn't
print runaway notices until Monday. She followed the North Star
and walked only at night. If it was cloudy she looked to the moss
growing on the shady side of the loblolly pine for orientation. She
talked with God. She drugged babies who cried. And if someone
got tired or wanted to turn back, she would hold a gun to their
head and say, *You'll be free or die a slave.*

＊

When I first meet Robin Grossinger it is eight o'clock on a
Wednesday morning and he is helping his son Leo out the door.
Leo is performing the national anthem at the A's game later, so he
has his schoolbag hanging from one shoulder, his clarinet tossed
over the other.

I walked from the BART station to Robin's house, set in the
borderlands between Berkeley and Oakland, passing hundreds
of flowering succulents as I went. Pygmy weed and aloe, jade and
spurge. Spires of tiny yellow blossoms thrust out of tremendous
plate-size purple aeoniums. It was unlike anything I had ever

seen. The rains that broke the five-year drought had coaxed nearly every plant into bloom. Though I knew Woolsey Street would have looked different last April, it was hard not to be optimistic. Perhaps San Francisco—thanks to its salt-worked past, powerful sense of environmental identity, and belief in the gospel of innovation—was going to be able to keep pace with sea level rise, at least in the short term. At least they were trying.

Nearly everyone working on wetlands restoration in the Bay Area has partnered with Robin at some point. He is a historical ecologist at the San Francisco Estuary Institute who uses old records—coastal survey maps, journal entries, photographs—to reconstruct an image of the bay as it once was. Folks like John Bourgeois use these renderings of the past to design and construct the landscapes of the future. On the Internet, Robin appears composed and curious, another member of that peculiar tribe that prefers to muck around in marsh mud while taking methane readings and chasing down mummichogs. In his author photo his ears protrude from under a sensible sun hat. But standing at the threshold of his egg-yolk-yellow bungalow, Robin seems somehow different. Less geek, more San Francisco chic. Arresting, even, with contemplative agate eyes. "He certainly casts a spell," a writer friend told me when I asked if Robin was similar to his quirky filmmaker sister Miranda July. And, in truth, I am a little unnerved. In the wild world of tidal wetlands restoration it is rare to run into a fellow traveler under the age of forty-five.

Together we walk through his eclectic house. Past the bright-orange IKEA chair made to mimic midcentury modern. "I have the same one at home!" I tell him with an unusual amount of pride. Past wall hangings, hand-carved wooden animals, and corkboards covered in photographs. Past a pile of shoes heaped near the stairs, and into the kitchen, where Robin offers me a cup of ginger tea. "I'm not too good at making coffee," he says.

"Oh, me neither," I reply before we move to the back porch to talk. We set our iPhones face down on the tiny tile table between us.

"My first task, back when I was a PhD student, was to research historical mapping of the bay's tidelands," Robin says. "To be honest, at the time it sounded totally boring." I nod in agreement, remembering how I felt when I realized writing about sea level rise would mean spending most of my time in salt marshes, a landscape that hadn't interested me then. "In order to figure out how we get the bay of today, I needed to figure out which species and ecosystems had been able to withstand the most alterations, which ones remained but were hidden, and which ones had completely disappeared. I started to think of it as a detective story."

"A detective story," I say. "I like that." To discover the direction of your own thinking in the course of mining the past—this is a practice historical ecologists and essayists share. The conclusion arrived at not in advance but through the process, by unearthing whatever is buried in the strata.

Already I wish our conversation could be longer. I feel as though I am in the presence of a kindred spirit, someone who understands just how difficult it is to make people care about ecosystems we spent the last couple hundred years lambasting and debasing. Someone who understands just how important these places are to the future of so many different species, ourselves included.

Robin tells me he spent years digging through piles of maps in the archives, mesmerized by their beauty and the clues they presented. Then he took copies of them out into the real world, in search of which elements persisted and which did not. "I remember when I realized that we used to have twenty-seven miles of sandy beach in the bay. That surprised me. But even more important was the realization that if those beaches aren't in our memories then they aren't in our lexicon, and if they aren't in our lexicon they also aren't in the palettes of all those landscape architects and naturalists attempting to bring back and buffer the estuary."

To an artist a palette is the oval of wood onto which she squeezes blobs of paint. It is a launchpad. A memory trigger. A marketplace. It is the intermediary stage in her creative process. If a color doesn't make it onto the palette, it cannot be chosen. Again I am reminded of the words of John Bear Mitchell, who said that the Penobscot people of Maine had been able to adapt to environmental change by narrowing their spiritual and physical palettes. But what if studying historic landscapes enables us to see and reintroduce some of the colors we thought we had lost?

Many of the remnant marshes in the Central Bay were once fronted by a beach, and that beach protected the wetland. It slowed erosion and provided nourishing sediment, helping the marshes make it into the present tense. "We're applying that historical insight to our interventions in the physical landscape today," Robin says. "In Marin County we're reusing sediment dredged from the shipping channels to build up a buffering edge along some of the most rapidly eroding marshes." Like John's project in the South Bay, this too is relying on recycled dirt to accentuate the marshland's inherent adaptive capacities. "It's *really* cool," Robin adds with a cluck.

Resilience then might mean keeping as many options available as possible. Putting as many blobs of paint onto the ecologists' palette as will fit. And experimenting now with various color combinations so we have a clearer idea in the future of what works where. "We've been defining resiliency in so many different ways for so long," Robin says. "Mostly this amounts to a hodgepodge of interventions that create temperature refuges for hummingbirds and velocity refuges for salmon. Specific solutions for specific animals in specific places. But if you understand the science of how these landscapes work, then you can't avoid thinking that bigger moves need to be made. We need to reconnect our ecosystems across landscapes. For example, we need to remove dams so that sediment trapped in reservoirs is actually coming down the streams and making it into the bay. That would increase the resilience of so many aspects of the system. It would put more silt into the water column, accelerating accretion rates that would help marshes gain ground."

What Robin is proposing is another shift toward "conserving the stage," as in Oregon's spotted owl territory, but with an even wider and much more purposeful lens. Instead of asking which locations or ecosystems certain species need to survive, "conserving the stage" means taking into account and working on behalf of the physical factors that foster biodiversity in the first place: soil types, hydrology, landform variation, and, above all else, topography. Robin's approach acknowledges that nature is dynamic and resilient. Instead of setting aside selected areas (a particular national park) or ecosystem types (wetlands refuges) as monuments to an idea of nature that is no longer tenable, he argues, we need to create arenas where evolution can continue to unfold.

Forty years ago, when the fight for the area's wetlands took shape, environmentalists often used the Endangered Species Act to earmark pockets of extant marshland for preservation. If they could find a single salt marsh harvest mouse rooting around in the pickleweed then they could make a case for saving that little postage stamp from development. Robin's proposal takes giant steps back from this conservation strategy. "We can't preserve the salt marsh harvest mouse if the salt marshes themselves cease to exist," he says, echoing John's statement that the fate of many endangered species is tied up in the future of the West Coast's tidal wetlands. "What science shows us is that we need to be infusing our work with an understanding of Big Nature, of landscape-scale change. And that includes different kinds of actions, involving multiple agencies, agencies that really don't have a responsibility to solve the whole problem. But if we can create partnerships across them—as we are doing here with Oro Loma, where wetlands restoration, flood resiliency, and sewage treatment are all being tackled together—then the results can be really profound."

"Oro Loma?" I say, repeating the enchanted words that I've already heard from John.

"Yeah, you *have* to make it out there," Robin says. "Think of it like this: If we can get Oro Loma to work, we will be, for the first time in history, reconnecting a kind of freshwater creek to the bay. And when we do that, we begin to feel reconnected to the deep nature of a place. Thinking—no, living—in this way is going to be better for us overall. We are going to have healthier, happier communities." He pauses and looks right at me. Do I buy it? Do I buy that sea level rise might not just be a catalyst for cataclysm, but might also pave the way for a massive, ultimately beneficial cultural transformation?

Yes, I want to say, yes. But who among us will get to live in the resilient, climate-ready cities we are designing along the water's edge?

❋

This is how Robert Moses made sure New York City stayed segregated: No one elected him to office, but he had more power than the mayor or the governor. More power than the two combined. At one moment in time he held twelve different positions. He was the head of the New York State Power Commission, the chairman of the Triborough Bridge and Tunnel Authority, the parks commissioner, and the city's construction coordinator. Today's New York, the city with the most segregated school system in

the country, is Moses's New York. He designed it to foster exclusion. During his reign as "master builder," as many often called
him, Moses constructed thirteen bridges and 416 miles of park-
ways. He purposely ran his network of superhighways, funnel-
ing white-collar workers between suburbia and the city, through
historically working-class neighborhoods. Places like the Bronx's
Mott Haven, Hunts Point, East Tremont, and Soundview. There
housing values plummeted, landlords set their buildings aflame,
and only the poorest of the poor remained. The destruction was
widespread and intentional, and its impact can still be felt today.
When Moses built the Long Island Expressway he designed each
of its 180 overpasses to be seven feet and seven inches tall. Too
low for a public bus to pass through. That way Jones Beach—
and each of its million-dollar bathhouses, which Moses also
erected—would remain, by perverse default, accessible only to
those with enough money to own a car. Of the 658 playgrounds
he built, the overwhelming majority were located in well-to-do
white neighborhoods. And when he did put a pool into Spanish
Harlem he kept the temperature "deliberately frigid" so no one
would want to swim.

✳

Walking away from Robin's house toward an afternoon of interviews on the University of California, Berkeley, campus, I think
back to my first morning in Alviso. Immediately I spotted two
giant glass cubes set high above the neighborhood at its west-
ernmost reach. This was the trucking and distribution center that
Richard Santos opposed. From where John and I stood, the cubes
appeared ten times taller than anything else, looming over the
single-story homes like the Emerald City over Oz. John used the
cubes and the Zanker landfill to orient himself. "Intel is there just

to the right, then beyond it is Yahoo!, and over by the pointed white roofs, that's *all* Google," he said.

As I walk I am thinking about Richard and all the other residents of Alviso, East Palo Alto, Redwood City, and Richmond. I am thinking that these places have long been relatively affordable because of their flooding problem, and how "conserving the stage" will likely introduce new stressors even as it removes others. I am thinking about the rapaciousness of capital, and how the nature refuges and flood resiliency projects going into these neighborhoods are likely to make them safer and prettier while also pricing out current residents. I am thinking that these people are sandwiched between rising tides on one side and Silicon Valley on the other, and that this position is not so different from the one that most tidelands species currently occupy.

I am thinking of something John said when we gazed out at this sea of shimmering buildings. "Facebook was the only company from Silicon Valley to contribute to wetlands restoration last year. They gave fifteen thousand dollars."

"I'll bet they can find that in their couch cushions," I responded. And we laughed but it wasn't all that funny.

Later I will discover that the year before Facebook made its meager donation, it paid $400 million for a fifty-six-acre flood-prone former industrial park adjacent to the Ravenswood Salt Ponds. This is land that will, according to the environmental impact report, require sea level rise mitigation, that will be subject to inundation with the sixteen inches of rise the state predicts by midcentury. Land that will, as it is developed, impede the potential inland migration of many of the species who currently live alongside it. Land that will benefit directly from the projects Facebook, by all practical measures, has yet to support.

I am thinking about how, when employees of Amazon and

Oracle and Intel and Facebook move into the working-class neighborhoods surrounding their disastrously low-lying tech campuses, they likely have the money to put their homes up on stilts. They have the money to pay for flood insurance. They have the money to protect themselves from sea level rise in a way that longtime residents cannot. And with each dollar that they sink into the sinking land, the more valuable it becomes and the more likely the local government is to fund the innovative, large-scale flood resiliency projects necessary to keep the waters out and the property taxes flowing in. I am thinking about Chris and Edison, and how the Isle de Jean Charles wasn't included in the Morganza to the Gulf protection plan. I am thinking it has as much to do with the perceived value of their land as it does with the cost of including them. Had the island been home to Google and not to members of the Biloxi-Chitimacha-Choctaw tribe, you can bet it would be surrounded by levees today. I am thinking of all the people in the Tanyard who flood when it rains and who flood when it storms and who can't afford to leave and can't afford to stay either. I am thinking about how each storm eats away a little bit more of the everything they've got.

I am thinking of Franca Costa, who was adamant when the Staten Island buyouts began: she would not be leaving her little "piece of paradise," since what she owed on her mortgage nearly eclipsed the offer the state had extended. But in the intervening years, her annual flood insurance has risen significantly and that cost has her thinking about selling. "I'm up to nineteen hundred dollars per year and I told myself I wouldn't go past two thousand. I don't have the money to lift the house and I don't have the money to insure it either," she said on the phone when I called her one day after work. "As far as I'm concerned I need to be out by June, when my premium increase will come around again."

I am thinking about the California Ridgway's rail and the salt marsh harvest mouse and the roseate spoonbill and the Caspian tern and the greater egret and the rufous hummingbird and the salt marsh sparrow and the red knot and the cypress and the black tupelo. And how horizontal levees will help, but are in no way going to provide sufficient space for all of them to move in. When I think about the different species I have encountered along the water's edge over the past five years, I know that we are all in this together. But I don't think that we have collectively come around to thinking in this way just yet. Though I hope—no, pray—that we will. Because I know that if we do nothing to address the ways in which sea level rise will deepen economic and social inequality while simultaneously displacing and potentially drowning half the species currently considered endangered, then we won't need a Robert Moses of sea level rise. The increased segregation and exclusion and extinction will happen all on their own.

I am thinking about how Silicon Valley and the tech industry and the innovative ethos of San Francisco are twenty-first-century versions of the same old get-rich-quick scheme, the same old narrative where the march of progress promises to transmute buried rocks into rocket fuel, deserts into cornfields, thin air into capital, stolen swamplands into private property. Facebook just constructed a 430,000-foot campus on former tidal wetlands that today rest only 1.6 feet above sea level. The company hired Frank Gehry to design it; it cost $195,824,452 to build. There are dozens of redwood stumps turned benches on the roof and a teepee-shaped swing. The meeting rooms have goofy names like "Seagulls over the Bay Are Bagels" and "Foods That End in Amburger." There is also a floor-to-ceiling mural of the word *WHY*. Before building, Facebook dumped 72,500 cubic yards of dirt onto the site to raise it above base flood elevation, and then it lifted the first floor even higher, perching the entire building above the ground on ten-foot-tall concrete stilts.

I am thinking that while Facebook purposefully, painstakingly lifted every single one of its new offices as protection from the first wave of future flooding, it didn't elevate much of the infrastructure the buildings depend upon. It didn't elevate the roadways or the storm pipes or the sewer system. When those flood, taxpayers will cover the expense. And when the salt marsh harvest mice living just east of the parking lot drown, few working at Facebook will know. Because what they do and who they are is not dependent upon the land where their company rests; if Facebook eventually relocates to higher ground, it will be exactly what it was before—a social networking platform that connects users globally, while disconnecting them from the physical setting where their lives take place. I am thinking about how the ability to move and remain unchanged is a privilege not shared equally by everyone and everything currently residing along the water's edge.

As I reach Berkeley, I am thinking about how tired I am of thinking in flawed time frames: the human time frame of roughly five generations—my grandparents, my parents, myself, and my children and grandchildren to come; the environmental impact report time frame, which in Menlo Park is only forty years; the insurance companies' time frame—the average length of a mortgage—which is, in the United States, thirty years; and the time frame of the developer, who typically tries to flip a building in five, or of the politician, who needs to get reelected every four. I am thinking about how tired I am of the surreal effort to adapt to climate change with rooftop gardens, stilts, swales, and hip wall art in a building that was purposely erected atop land that already floods. I am thinking that the belief that we can design our way out of this is part of the same set of addictions we must learn to give up. I am thinking about justice, and what it might look like if we thought of sea level rise as an opportunity to mend our relationship with the land and with each other.

This is how we prepare to retreat: Too many times I have been told that there will never be enough money in the federal coffers to relocate everyone away from the risk of rising tides. This is true until we decide to make it untrue. Lately I have been thinking we should mimic California's Measure AA and institute a nationwide property tax of one penny per square foot per year. Call it the "Get out the Way Tax" or the "We Refuse to Rearrange the Deck Chairs on the *Titanic* Tax" or the "Saving Half the Endangered Species Tax" or the "Let's Not Make Mars the Backup Plan Tax" or the "We Resolve to Act with Empathy Tax." In paying it we would recognize that owning property is always both a privilege and a risk, and that we are each of us exposed at any given moment. In most communities plagued by chronic flooding, residents don't leave for one of two reasons: they can't afford to relocate, or they feel forced. A managed retreat tax would begin to help us overcome the first hurdle and also possibly the second; if the process is widespread it won't feel as unfair. It would allow us to collectively practice walking away from something we desire. Maybe we could call it the "Seas Are Rising and So Are We Tax."

*

I visit Oro Loma. It is breathtakingly beautiful in its understated way. Behind the horizontal levee is a small freshwater treatment wetland, where partly cleaned wastewater filters through cattails and bulrushes to break down contaminants and remove nutrients that cause excessive algal growth in the bay. Then it seeps into the wide levee, through willows and creeping wild rye, Baltic rush and basket sedge, western ragweed and California blackberry, all laid down in rows. These plants continue the work the treatment wetland began, sucking up the nitrates and phosphorus that remain.

Over the past thirty years many of the region's saltwater marshes have been restored, but most have little connection with the creek mouths, moist meadows, and willow groves that once served as transitional zones, or ecotones, between tidal habitats and terrestrial ones. Oro Loma is an attempt to engineer some of those missing elements back into the landscape while doing double duty, both cleaning effluent and providing much-needed flood protection. And yet as I look out over this wide, man-made levee, my mind is pulled in two directions at once.

Is Oro Loma too similar to the massive infrastructural missteps we have made in the past? A seemingly easy solution that might compound the problem in unforeseen ways? Or is this five-hundred-foot-wide earthen berm ultimately trivial, even if one existed in every rehabilitated wetland from San Jose to Marin County—its meager size ill suited to the immense changes that sea level rise has already set in motion? I circle the project two, three, four times on foot, unable to decide. Too big or too small? Too hubristic or too narrow sighted? At least it is not another seawall or another set of stilts. At least it is attempting to mimic the natural defenses we have spent the past two centuries paving over. At least it is designed to benefit species other than ours. This counts for something.

The air is musty with the smell of sewage, and salt water laps at the concrete retaining wall separating the project from the bay. Many more tests need to be run before the sanitation department can release the treated wastewater into the largest estuary on the West Coast. For now the system remains closed. An experiment that many hope will turn into an adaptation strategy. Eventually I thank the project manager for the tour and drive to a local Vietnamese restaurant for a bowl of *bún chả* and a cup of strong coffee.

On my way out of town, I stop at the San Francisco Estuary Institute to speak with Jeremy Lowe, a senior environmental scientist for the Resilient Landscapes Program that Robin runs and the person behind the horizontal levee concept. Jeremy is a coastal geomorphologist who has been working in climate change adaptation for over thirty years. He has designed floating floodgates for Venice and tried to minimize coastal erosion at the Hong Kong airport. But for the last decade or so he has focused on nature-based sea level rise readiness strategies for the West Coast. He has a mild manner and a fabulous British accent. And not more than five minutes into our meeting, he surprises me by saying what no one else working in wetlands restoration and resiliency will: That Oro Loma and the South Bay Salt Pond Restoration Project and the innovative uses of dredged materials aren't solutions in and of themselves. That they aren't going to reverse the tide and they aren't going to solve the problem of equity. That there is no silver bullet.

We are sitting together, looking at a map of the bay that Jeremy's institute recently published. It inventories all the different kinds of resiliency projects taking place, from wetlands restoration to horizontal levees and beach replenishment. "What we're doing is buying time, some buffer, in which to wrap our heads around the fact that—in the grand scheme of things—this isn't going to work," Jeremy says. "We're going to have to move infrastructure; we're going to have to move people. Lots of them. And in the meantime, if we can make our marshes more resilient to buy us all—humans, plants, and animals—some breathing room in which to figure out how to retreat responsibly, then let's do that."

We talk for a while longer, then I walk out of his office, down the spiral staircase to the lobby, and outside. The early-evening breeze turns the leaves of the cottonwood. Costco is to my right,

Ice Chamber Athletic Performance Training to my left. In front of me a snowy egret stands knee-deep in the bay. Jeremy's words ring in my ears: "This isn't going to work." None of it: not the horizontal levees or the sediment slurries, not the seawalls or the living dunes, not the raising of homes and streets or the pumping out of salt water, not the flood insurance reform or the visions of twenty-second-century Venice. The water will come and at some point we are going to have to admit a kind of defeat. That nothing important is ever easy or quick. That real resiliency might mean letting go of our image of the coastline, learning to leave the very places we have long considered necessary to our survival.

Organized retreat is one of the few adaptive strategies that feels appropriately humble and, at the same time, acknowledges the scale of the threat. It is the only one that calls for everyone living on the lowest-lying land along the water's edge, those who can afford to lift their homes and those who cannot, to participate. Because in order for retreat to make long-term financial sense, in order for the gains to outweigh the losses, we all have to move in together. Only then will we quit paying to repeatedly repair roadways and electric lines, storm-water infrastructure and sewer pipes, schools and bus stations and wastewater treatment plants and hundreds of thousands of residential homes that flood when it rains—and when it storms—and sometimes, now, even when the sun is shining. It is this, the radically egalitarian nature of retreat, that interests me most. Because individual vulnerability isn't arrived at by chance, it's a product of the same plunderous system that got us into this mess. We must learn to retreat because, in the end, there are other living beings far more vulnerable even than the residents of East Palo Alto and Alviso. We must learn to retreat because it is the sole adaptive strategy that will give this numinous egret the chance to move in too.

I sit at a picnic table on the edge of the bay for a long time, thinking of something I saw the previous evening in Alviso. While working my way through the crosshatch of levees that surround the little hamlet, I inadvertently ran into the old marina. At first it seemed like nothing more than a parking lot, a boat ramp, and a narrow slough. But when I marched over to read a "Fish Smart" sign illustrating which species carry dangerous traces of heavy metals in their bodies, I stumbled upon half a dozen wooden boardwalks connecting the gravel lot to the salt ponds.

The sun was setting and teenagers were strolling hand in hand out into the arching bull grasses and the dying light. To reach the windblown rushes and the hundreds of migrating birds lofted above, every visitor has to walk through a house-shaped wooden cutout. There are six in all and each one is painted pale yellow. They almost make the tidal marsh look like a backdrop for Our Town, quaint and undisturbed by the larger workings of the world.

I watched four couples pass through the gates of this Potemkin village before I decided to try it myself, expecting to find a sign explaining the cutouts' meaning or purpose. But there wasn't one. I strolled from home to home, searching for a way to read the installation as intended, but I never did uncover the key. What I realized, however, is that all of the homes look exactly the same from both directions. The sameness made me stop at each threshold. There were the marina and the towns of Alviso and San Jose and San Francisco beyond—and there, just as present, needing to be reckoned with, there were the wetlands, with their egrets and avocets and salt marsh mice. The installation gave equal weight to both tidal and terrestrial land.

Soon darkness would descend and I would head back toward the trailer I had rented on Airbnb. But first I sat very still on the edge of that marsh and listened as the sound of human chatter and bicycle bells mingled with a half dozen different birdcalls. The Wilson's warblers were in from Guatemala, the long-billed dowitcher from Baja. Soon the double-crested cormorants, with their deep guttural grunts, would head north along with the Swainson's thrushes, who might just make it all the way to the Andrews. There I was surrounded by members of my scattered tribe. By those who, like me, return to these often-overlooked places out of necessity, to touch—if only briefly—a living system so complex the sole word to describe it is divine. On the boardwalk I bumped into Lupe, who lives next to my rental on State Street and migrated from Mexico over fifty years ago. A few minutes later a mother and her daughter walked past speaking Chinese. And just before I left, a young couple jogged by, chatting about computer codes. Though I wanted to eavesdrop on all of it, in truth I understood little. A birdcall I couldn't identify sounded high overhead and I thought that—as is always the case in tidal wetlands—this soggy acre was a surprisingly cosmopolitan spot. The honeyed stands of bull grass swayed, turning green, then yellow, then back again in the last light of the day. From there I could see that there is no real difference separating one side of the plywood house from the other; all of it is home, at least until it isn't.

Afterword

Listening at the Water's Edge

I STILL RETURN TO OAKWOOD BEACH SOMETIMES. MOST of the homes are gone by now, the land where they once stood repopulated with fennel and wild parsnip and Queen Anne's lace. Deer graze on the native grasses and other wild plants the city seeded; Canadian geese waddle down the roads where kids used to play basketball.

I always walk along the disintegrating berm and down what was once Fox Beach Avenue. I walk past the place where Nicole Montalto's house stood, past Joe Tirone's lot and Pedro Correa's. On my most recent visit I noticed Franca Costa's home had finally been demolished, too, so I emailed her to ask for an address where I might send a copy of this book.

Franca, I learned, had moved to Florida. "I feel safe in my new home," she wrote. "When severe weather strikes I am no longer nervous. The river is close but not very close, and way downhill from here!" Choosing to leave took Franca longer than it did others; it was a decision she had to reach on her own, the journey toward her departure unique. However, moving away from risk would have been impossible without the financial assistance of the buyout program. She told me she got just enough money to relocate to Daytona Beach, where much of her family lives. "I look back and am surprised it took me so long to leave. Now I no longer want to live by the water. Funny how we change over time."

On my most recent trip to New York City, Joe invited me to lunch with Frank Moszczynski and John Toto, the two other leaders of Staten Island's successful buyout movements. We ate fish sandwiches at the family restaurant John owns and chatted about what had changed in the five years since the buyouts began and also what hadn't.

"I didn't move far," Frank said. "Just up the hill and out of the floodplain."

"Me neither. I still have this place here," John added, touching the wooden tabletop.

They were all living close to the homes they'd lost to the storm. They still went to the same butchers and grocery stores, and hung out with the same people on the weekends. Joe was still a local real estate broker and Frank still worked construction with the city's recovery program. Sometimes they went down to the pier near Oakwood to fish. Their community network remained largely intact; what had changed was their immediate vulnerability to flooding.

I asked how many people from the buyouts remained on Staten Island, and Joe said, without missing a beat, "About eighty percent." At first, I figured his number was exaggerated, that more residents had been like Franca and moved far away—to Florida, upstate, or elsewhere. But later that month, during a preparatory conversation for a panel at the National Adaptation Forum, one of my fellow proposed presenters shared her findings that 79 percent of participants in the Staten Island buyouts had indeed remained within a five-mile radius of their former homes. Many of them had moved to areas with lower income thresholds, but all were able to maintain the bonds they had built over decades with their fellow Oakwood residents.

When we talk about buyouts we tend to focus on the losses they perpetuate: the loss of community, the loss of property taxes, the loss of a homeplace. But like so much else in climate discourse, this is a discussion that plays out far from the water's edge, and it frames those impacted by higher tides and stronger storms solely as victims. When I listen to the stories rising from Staten Island's eastern shore, I hear a counter-narrative, one of transformation and of hope. Oakwood's former residents wanted to leave a legacy, to be part of something larger than themselves, and to aid adjacent neighborhoods in the process. By joining forces and moving away from risk these people became agents of change, not just flood victims or refugees. The buyout, surprisingly, brought them closer together.

✻

"We tell ourselves stories in order to live," Joan Didion writes at the opening of "The White Album," her famous essay that tries and fails to make sense of how the idealism of the 1960s and California's "golden dream" gave way to, was consumed by, the Manson murders and a kind of unrelenting cynicism. Didion says that storytelling works as a sense-making practice, at least until it doesn't. There are moments and phenomena that test and ultimately rend our ability to arrive at a narrative line.

Certainly climate change is among them. And yet we keep trying to tell this story in a straightforward manner, with conventional narrative techniques and news reporting. Climate change has entered into our contemporary culture as a never-ending set of "record-breaking" statistics—record-breaking storms, record-breaking heat waves, record-breaking rain—each successive extreme smashing the previous record-breaking record. Since I finished writing *Rising*, five different supercharged hurricanes— Harvey, Maria, Irma, Florence, and Michael—have unsettled coastal communities all around the United States. Meanwhile wildfires rage from California to Washington and heat waves kill those isolated deep within our urban centers.

Many ask if this is the new normal. But there is nothing normal about a single species setting in motion such a significant geologic shift, the largest in our history. And every time we write about climate change with the same tired vernacular, we dull readers to the dynamism at the heart of such transformation. Put another way: we have plenty of climate change news, but this reporting tells the story straight, and so we think the conclusion is foregone. By describing climate change in this manner we steal some of its mystery, what Amitav Ghosh calls its "improbability" or "uncanniness." In so doing, we also steal from ourselves the possibility that we might be transformed, and not just for the worse, by this disruptive force.

We need only look to Staten Island for evidence of that electric possibility. When communities long made vulnerable by existing structural inequalities are also directly impacted by climate change, it can awaken an awareness of vulnerability, and also an awareness that this vulnerability is shared. This realization brought the residents of Oakwood Beach together: demanding access to one of the most progressive sea level rise adaptation techniques we have, and, at an even more basic level, inspiring them to raise their voices and regain control over their community's destiny.

The question remains: How to tell this story so that it becomes more than elegy alone, both a record of these uncanny times and also a rallying cry? The longer I steeped myself in the early impacts of sea level rise on coastal residents around the country, the more profoundly I understood that these people needed to tell their own stories. That no amount of mediation on my part could make their words more powerful. So I recorded my interviews, transcribed them, and pared them down, with the residents' help, until what had been a conversation over a glass of sweet tea became a testimony bearing witness to the earth and its changes. In this way I was able to invite readers into the experience that most transformed me while at work on *Rising*: that of sitting in a stranger's living room and listening to how the water had worked its way in, up the driveway and under the door. With it came the sinking knowledge that soon the floor would buckle, if it hadn't already; that the Sheetrock would need to go; that the family photos would warp. That the home this person built, made a life inside of, would come undone.

When residents spoke of the way flooding was dismantling their homes, the words they used were, for me, a kind of poetry. Their language cut to the marrow of lived experience. Nicole said, "I went down to the basement and began screaming. I was hoping that I would hear him but at the same time I wasn't." That line has a music to it. On the surface it is contradictory, yet both sides are true, and the music is what holds them together. I suppose that is what I was searching for—the stories that sang; the stories that will bind us to one another as our homeplaces change irrevocably right under our feet. The stories that, as Ghosh writes, acknowledge "the magnitude and interconnectedness of the transformations that are now underway."

Along with residents' knowledge of just how fundamentally the coastline was changing and how vulnerable and exposed they were (and, in many cases, still are) came a heightened awareness of the land where they had long made their lives. Here I was, listening to people from all across the political spectrum talk about how flooding taught them to pay close attention to the more-than-human world of which they are a part. Since the rise of scientific rationalism, our particular brand of western knowledge has lulled us into thinking that we are separate from nature: if we can design our way out of feeling the cold at the heart of winter, we can also simply raise our communities above the storm. But now climate change is calling us to attention, drawing us to the water's edge to ask with wonderment and fear whether there is, or ever really was, something that separates us from our environment.

*

Recently New York City announced plans to build a four-and-a-half-mile-long seawall from Fort Wadsworth to Oakwood Beach. The project, designed by the Army Corps of Engineers, will augment the existing levee system, adding fill to the berms that failed during Sandy. During my lunch with the buyout leaders, I asked what they made of the city's proposal.

"It's going to be a little higher than what we have now. But we also know that seas are rising," Joe said, shaking his head. "If you ask me, that seawall is only a temporary solution, and it's going to make residents feel safer than they are."

When I had started visiting Oakwood five years earlier, next to no one would utter the words "sea level rise." Folks would tell me that their flooding problems were getting worse, but that they weren't sure of the cause. My conversation with the buyout leaders is not representative of the opinions of all of Oakwood's former residents, of course. It does illustrate, however, how talking about climate change has become less taboo in this Republican enclave.

Candis Callison writes about the gulf between local and national vernaculars of climate change in her book *How Climate Change Comes to Matter: The Communal Life of Facts*. Recalling her time in an indigenous fishing village in Alaska, she writes, "I experienced not an explicit questioning of climate change but a flat-out rejection of it as a term that described what direct experience with climatic changes feels like and how it is that such changes are understood and discussed." According to Callison, the Inuit will be the first to tell you that winters are warmer, that the sea ice is breaking up, and that many precious species are disappearing. They just don't like the term "climate change," and who could blame them? "Climate change" sounds like a phrase that a government agency might employ to justify new fishing regulations or a phenomenon a scholar from outside the community might study. The words themselves ring as too political, too cold, too loaded to map onto everyday life. Moreover, they suggest a limited set of pathways forward, pathways that are frequently dreamed up elsewhere and do not take residents' lived experience and local knowledge into consideration.

In Oakwood Beach, citizens came together. They created signs and petitions and a publicity campaign. They fought for a just recovery from Sandy, and when they won the right to adapt to the storm in the way that felt most in line with the wishes of their community, when their ideas were honored, many acknowledged that careless development practices combined with human-driven climate change had together caused their community to flood. "We tell ourselves stories in order to live": my work on this book regularly reminded me that the opportunity to speak about your own shifting relationship with the environment and to have those stories heard is something that ought to be extended equally to everyone—regardless of the specific language we choose to use—and yet all too often is not.

I used to think of testimonial storytelling as an opportunity to "give voice" to those communities long kept at arm's length from official climate change discourse. But over time that description came to ring untrue. I was not giving voice to coastal residents. They had voices: powerful, distinct, poetic voices. Instead I was giving them a microphone.

That I might have a microphone in the first place has much to do with my privilege as a white woman and how that intersects with the history of environmentalism in this country. As the child of lapsed hippies in New England, I spent my summers in Maine's iconic national park, Acadia. There I scrambled up Cadillac Mountain—named after a French explorer—to watch the sun rise. I read *Walden* and *A Sand County Almanac* and, later, *Desert Solitaire*, *My First Summer in the Sierra*, and *South*. The more I read, the more familiar these extraordinary tales of adventure and discovery became, until I started to see myself in their white pages. Which is to say: I inherited not only a love of wandering in the mountains but also these stories, which told me that individualism, risk taking, and a heady mixture of paying attention to the natural world and can-do gumption on the frontier (be it here on the continent, at the poles, or in any other host of remote "undiscovered" environments) has made "us" Americans what we are. But there are so very many that "us" leaves out—from the tens of millions of indigenous peoples killed through the genocidal colonial project to those whose ancestral roots connect them to Africa and Asia and Native America. Too often today that "us" also excludes those without the means to escape to the wilderness on the weekend.

Looking back, it is easy enough to identify these gaps in environmental discourse. Decidedly more difficult is the question of what might make this conversation, this movement, whole moving forward. A few months prior to my most recent Oakwood visit, I heard Zoe Nyssa of Purdue University give an excellent presentation in which she traced how the words we use to describe the environment and our relationship to it have shifted over time. In the 1980s we typically described the environment with words like "landscape," "sustainable," "natural," and "endangered." People in relationship to the environment were "man," "youth," and "citizen." But much has changed in the intervening three decades. Today, Zoe's research shows, we describe both people *and* the environment with the word "resilient." Our language has narrowed and refocused on empowerment—but precisely who is empowered and by what mechanisms differs radically depending on where you stand.

Because what do we mean exactly when we call someone or something "resilient"? In Manhattan, sea level rise resilience means—in the case of the plan that won the Rebuild by Design competition after Sandy—"landscaped berms with sea grass and habitat supports, levees that double as skate parks and amphitheaters, sea walls that support pop-up cafés and passive recreation." I am not inherently against these things, but in lower-income communities of color in New Orleans, for instance, "resilience" can mean something else entirely. Ten years after Katrina, only 37 percent of the residents of the Lower Ninth Ward had returned. That neighborhood once had one of the highest rates of black homeownership in the city—but property values had also long been suppressed by redlining, which meant that recovery funding based on the home's market value frequently did not cover the basic cost of rebuilding. In this case, resilience meant not getting enough money to recover in place and having to start over somewhere else. Some residents would head to Houston, only to be displaced by another unprecedented storm, Hurricane

Harvey. For those who remained, resilience might mean living in a community where abandoned homes outnumbered occupied properties and where municipal services were slow to return. The Lower Ninth Ward was the last neighborhood in New Orleans to have its electricity supply restored and for months it was unclear whether the community's public schools would reopen at all.

Tracie Washington of the Louisiana Justice Institute spoke at the Rockefeller Foundation's launch of the "Resilient New Orleans" initiative, saying, "Stop calling me resilient, because every time you say, 'Oh, they're resilient' that means you can do something else to me. I am not resilient." These words were printed on posters and hung throughout the city by local activists to mark the ten-year anniversary of the storm. The proliferation of this straightforward statement, both in many of the city's lower-income communities of color and around the country, speaks directly to the fact that New Orleans's "revival" (as it is described in the pages of *Fortune*) has not benefited everyone equally. Climate change impacts and environmental injustice often overlap, further deepening the distances between those communities that will receive institutional support to weather the storms to come and those that will not.

In Katrina's wet wake, the city infamously proposed to turn many of its lowest-income, lowest-lying communities—the Lower Ninth Ward included—into open space that might serve as a buffer in future storms. It published its plan on the front page of the *Times-Picayune*, with a map that marked, with green dots, residential areas that might be remade as parks. Those who lived in these communities were furious, and rightly so: no one had consulted them. "This map was perceived as showing how more powerful 'others' were trying to dictate where people should and should not live," says Liz Koslov, a professor in the department of urban planning and the Institute of the Environment and Sustainability at UCLA. In New Orleans, residents dug in their heels. They were not going to cede the flood-prone land they had been forced to occupy by discriminatory lending practices, the places they had made their own through years of attention and care, just because the city suddenly decided that it wanted that land back.

The proposed recovery strategy would have required buyouts similar to those that took place in Staten Island, and yet in the Big Easy the idea was quickly dropped. What makes managed retreat successful in some places and the root of great derision in others? I suspect it has everything to do with whether or not residents author the story of their own recovery. In Oakwood, Joe proposed the idea to his neighbors. Together they made maps by hand of the potential buyout zones, highlighting the spaces they no longer wanted to live. In New Orleans, the "green dot" map was a product of then-Mayor Ray Nagin's Bring New Orleans Back Commission; it was created outside of the communities it sought to dismantle.

When my nonfiction students jump into a new writing project, I ask that they immerse themselves in the tradition of which they wish to be a part and to try to note what is missing. Whose voices or perspectives have been left out? For too long, environmental concerns in the United States have largely been the domain of those who can afford to visit nature in their free time, who have the privilege to choose where and how they want to live. But the reality is that many living on climate change's front lines are low- to working-class people and communities of color, whose relationships with the more-than-human world regularly go unaccounted for in the "official story" of environmentalism we tell in this country.

During a *Rising* event at a San Francisco bookstore, a middle-aged white man in a brown leather jacket and an Earth Day ball cap asked, "But what do nonbelievers say when you show them the charts that illustrate temperature change over time? How can they deny the climate science?" I was discomfited, but not surprised, by his question, given the particularly divisive moment we occupy. "I don't go into my interviews with charts that illustrate how the earth is warming," I responded. "I leave my discourse at the door and instead ask residents to tell me their flood stories. I am there to listen."

It sounds deceptively simple, and maybe it is. But when a conversation has long been dominated by a select few, listening can be a surprisingly potent act, upending historic power imbalances. As the tides get higher and storms stronger, those long exposed to flooding have precious knowledge that the rest of us do not. They know how to read the land and, where possible, how to identify pathways forward.

I suspect fewer people will resist the term "climate change" not just when the floodwaters arrive at their doorsteps but also when they get to tell their stories and sense that they have been heard. When the words they alight upon, excavate, and share become part of the vernacular we use to describe these uncanny and improbable days, *that* is when this phenomenon will become more than a catalyst for cataclysm. As the language around climate change loosens, becomes more democratic, our ability to seize this moment as an opportunity for coalition building, especially among those long made vulnerable by other structural problems, will grow. Perhaps together we will make the ever more popular protest chant come true: "The seas are rising and so are we."

ACKNOWLEDGMENTS

Books are strange beasts, born exactly where or when or by what means it is hard to say. So many people came together to bring *Rising* into the world, and to each of you I am deeply grateful.

First and foremost, I wish to thank those whose voices fill these pages alongside my own. I think of you as the chorus rising from the water's edge, showing us the way forward. My debt to you will never be fully repaid. Thank you for agreeing to be part of *Rising*.

Chris Brunet, you welcomed me onto the island and into your home. In the five years between then and now, atop that salty land, a rich friendship has grown.

Nicole Montalto, when you told me the story of the storm that took your father's life, I knew that I needed the voices of those who are being affected by flooding to enter into this text unmediated. Your courage and strength taught me to listen. This was an invaluable lesson.

Marilynn Wiggins, you helped me begin to see the way that just the threat of flood insurance reform is already reshaping our most vulnerable communities. You helped me to also understand how danger is always felt in the body first.

Laura Sewall, your friendship made me feel right at home on the farthest edge of the marsh and in the world as it comes undone.

Dan Kipnis, thank you for your courage in saying what so few others will, that the time to leave the homes we have built along the coast is now.

Richard Santos, your candor and your love of your community are a great example to us all.

Thank you to the many different scientists who allowed me to tag along, to ask a lot of silly questions, and then, when I didn't understand the answer the first time, to ask again. In particular, Beverly Johnson, Pete Frezza, Hal Wanless, Ben Strauss, Cameron McCormick, Sarah Frey, Thomas Doyle, Brett Hartyl, Steve Akers, John Bourgeois, Robin Grossinger, Jeremy Lowe, and Tiffany Troxler and her team. Thank you also to the hundreds of scientists whose research built this book's backbone.

Thank you to my many local guides: Joe Tirone, Franca Costa, Patti Snyder, and Loisann Kelly in Oakwood Beach; Albert Naquin and Edison Dardar on the Isle de Jean Charles; Nicole Hernandez Hammer in Shorecrest; Alvin Turner and Marilyn Montgomery in Pensacola; Bryan Doyle, Frederick Swanson, and Charles Goodrich at the H. J. Andrews Experimental Forest; and Len Materman and Susan Schwartzenberg in San Francisco.

Various institutions have supported *Rising* with the gift of time and space to work and, in a few lucky instances, money. I owe a great debt to the Andrew W. Mellon Foundation, the Howard Foundation at Brown University, the Metcalf Institute for Marine & Environmental Reporting, the Society of Environmental Journalists, the Bates Faculty Development Fund, and the National Association of Science Writers. Additionally, both Oregon State's Spring Creek Project and Playa provided refuges where I was able to write while also wringing my body out in the wildernesses that abut both places.

Thanks to *Guernica* for publishing "The Marsh at the End of the World," and to the following magazines, journals, and newspapers for publishing portions, sometimes in different forms, from the following chapters: *The Atlantic* ("Looking Backward and Forward in Time"), *Creative Nonfiction* ("The Password"), the *Dark Mountain Project* and the *Global Oneness Project* ("Persimmons"), *The Guardian* and *Pacific Standard* ("Pulse"), the *Los Angeles Review of Books* ("Listening at the Water's Edge"), and the *New Republic* ("Divining Rod").Thank you to the entire Milkweed team for the care and attention you have shown to *Rising*, especially Daniel Slager, Joanna Demkiewicz, Abby Travis, Jordan Bascom, Meagan Bachmayer, and Mary Austin Speaker. Finally, my deepest thanks to Joey McGarvey, my exacting editor. Always, you asked the toughest questions, the ones that made *Rising* into something well wrought. And when I thought these sentences were complete you returned to them, again, generously and with attention. Like a squeeze of lemon juice in soup, your wisdom and wordsmithing are what make this book shine.

Julia Lord, you are both my agent and a guiding light. You are kind, wise, and steadfastly devoted to your authors' success, and I cherish our friendship and the many candlelit dinners shared in your family's company on West Ninth Street.

Katie Ford, you taught me how to write poetry. For this I owe you everything.

Katie Towler, Craig Childs, Rick Carey, Bob Begiebing, Gretchen Legler: fabulous mentors all. You made my years at Southern New Hampshire University meaningful and joyous.

Liz Koslov and Rebecca Elliott, whose dissertation work has regularly worked its way into these pages; our many conversations about retreat and flood insurance have advanced my thinking in all the right ways. I think of you both as peers, and consider it an honor.

Jane Costlow and Meera Subramanian, you two are kindred spirits, and your time with this book, your countless readings and conversations about environmental writing have profoundly shaped *Rising*. I cherish your company and camaraderie.

Kristen Stern and Janet Bourne and all of the other fine folks who passed through the Bates Junior Faculty Writing Group, thank you for reading early drafts of the essays that would eventually become this book and for offering your thoughtful feedback. Our weekly meetings were a wonderful reminder that writing is communal work.

Rob Farnsworth, Jess Anthony, Danny Danforth, and the rest of the Bates College community: I wrote much of this book in your blessed company, and I can think of no more encouraging or welcoming group of scholars. I knew when I first set foot on campus that my years at Bates would be very special; I was not mistaken.

Thank you to my students at the College of Staten Island, Bates College, and Brown University. It is an honor to explore creative writing in your company. Full stop. Your energy and enthusiasm and hunger feed my own.

Elise Bonner and Lauren Lanahan, my dear traveling companions, who walk alongside me through life: this journey would be so much less without the gift of your friendship. And to Anna Tellez, Zoey Laskaris, and Fiona Gladstone, our time together on mountaintops pumps both joy and solace into my days. Thank you to Rebecca Herman and David Bowles, friends since we first came together on the far side of Acadia.

Carole and Bill, my second parents: your support and love have buoyed this ship through both wild weather and blissful calm. And to Suzanne Lecht, Leah Brecher-Cohn, and Faviana Olivier, you wise women taught me, from a very young age, to be both strong and willing to lose yourself in song, always.

John and Martha, my parents, the love you wrapped me in from the very start is a protective shield and a source of energy. From it the endless sense of possibility pulses forth. Thank you for that and all I cannot name.

Felipe, my beloved. That I get to wake up next to you and that thousand-watt joy in your heart is, in truth, all I have ever wanted.

The rest of my gratitude is for the sea's brilliant iris and the sudden stone stillness of lake water in late summer, for the hum in my body that drowns out the mind when I walk back down the mountain.

NOTES

Epigraph

Taken from an interview with John Bear Mitchell conducted by two of my students, Hadley Moreau and Abigail Horrisberger, as part of a class project titled "Climate Change and the Stories We Tell," April 30, 2016. Together we transcribed and edited the interview. You can find a longer version of it online at http://www. bates.edu/climatechange.

The Password

1   *from Rhode Island back over two thousand years* For more on the Indigenous history of Rhode Island, in particular its relationship to the bay, check out Sarah Schumann, *Rhode Island's Shellfish Heritage: An Ecological History* (Rhode Island Sea Grant, 2015).

2   *It is Native American in origin* A.J. Hendershott, "The Other Swamp Tree," *Missouri Conservationist*, November 2001, https://www.xplormo.org/conmag/2001/11/other-swamp-tree.

2   *"Sometimes a key arrives before the lock"* From Rebecca Solnit, *The Faraway Nearby* (New York: Penguin Books, 2014).

2   *No state (save Maryland)* "States and Territories Working on Coastal Management," Office for Coastal Management, National Oceanic and Atmospheric Administration, October 14, 2016, https://coast.noaa.gov/czm/mystate/.

2   *15 percent of Rhode Island is classified as wetlands* This is a tricky number to pin down. I averaged three reports, placing more emphasis on those with finer analyses. The

first, and least robust, was published by the Natural Resources Conservation Service, run by the United States Department of Agriculture, titled "The Status and Recent Trends of Wetlands in the United States" (2010), which projects that 30 percent of the state is classified as wetlands. Meanwhile the Rhode Island Department of Environmental Management suggests that roughly 13 percent of the state is wetlands. See chapter two of their 2015 "Rhode Island's Fish and Wildlife Habitat" report, http://www.dem.ri.gov/programs/bnatres/fishwild/swap/chap2draft.pdf. Finally, the US Department of the Interior and the Fish and Wildlife Service projects that 10 percent of the state's land surface is represented by wetlands. See the Rhode Island entry in their 1989 report titled "National Wetlands Inventory," https://www.fws.gov/wetlands/Documents%5CWetlands-of-Rhode-Island.pdf.

2   *of that 15 percent, roughly an eighth is tidal* This number is derived from "Rhode Island's Fish and Wildlife Habitat," which contains the most precise estimate of total estuarine habitats, including tidal flats, salt marsh, brackish marsh, and various types of marine zone grasses.

2   *Over the past two hundred years* To learn more about Rhode Island's salt marshes of the past and future, check out the "Rhode Island Sea Level Affecting Marshes Model (SLAMM)" project published by the National Oceanic and Atmospheric Administration and its partners in 2015, http://www.crmc.ri.gov/maps/maps__slamm.html.

5   *There is a word coastal landscape architects* Here I am thinking of Azimuth Land Craft's "Climate Chronograph," the design that recently won the National Parks Service's *Memorial for the Future* design competition.

5   *the word especially refers* Random House Dictionary, s.v. "rampike," accessed March 19, 2018, http://www.dictionary.com/browse/rampike.

6   *"naming is the beginning of justice"* This heartening piece of writing is called "Speaking of Nature" and it appeared in the March/April 2017 issue of *Orion* magazine, at https://orionmagazine.org/article/speaking-of-nature/.

6   *Sometime during the last half century* Commonly it is thought that sea level rise began to accelerate in the early 1970s and ramped up even more significantly at the very end of the twentieth century. Also see Thomas W. Doyle et al., "Assessing the Impact of Tidal Flooding and Salinity on Long-term Growth of Baldcypress under Changing Climate and Riverflow," in *Ecology of Tidal Freshwater Forested Wetlands of the Southeastern United States* (Dordrecht, Netherlands: Springer, 2008).

6   *a recent bird census* This information came from an interview with Kenny Raposa, head researcher at the Narragansett Bay National Estuarine Research Reserve on Prudence Island, on September 10, 2015.

6   *The oldest living black tupelo* Taken from an interview with Cameron McCormick at the Audubon Society in Bristol on August 14, 2015, as are the quotes from Cameron in this chapter.

6   *killing off approximately one-third of Europe* As you can imagine, estimating the to-
     tal population of Europe and the number of people killed by the plague during
     the fourteenth century is difficult. The "one-third" statistic comes from the
     lower boundary of a range in Mike Ibeji, "Black Death," BBC, March 10, 2011,
     http://www.bbc.co.uk/history/british/middle__ages/black__01.shtml, while the
     Centers for Disease Control and Prevention estimate 60 percent of the populace
     was killed, https://www.cdc.gov/plague/history/index.html.

6   *And the salt marsh sparrow's* Alex Kuffner, "Drowning Marshes: Where Does a
     Species Go When the Nursery Floods," *Providence Journal*, April 9, 2016, http://
     www.providencejournal.com/article/20160409/NEWS/160409256.

6   *Of the fourteen hundred endangered or threatened species* This number comes from the
     US Fish & Wildlife Service's online listing of all endangered and threatened spe-
     cies, which is updated yearly. In 2015, 2016, and 2017, over one hundred species
     were added to the list. To access the list, visit: https://ecos.fws.gov/ecp/.

6   *over half are wetland dependent* For more on the overlap between endangered and
     threatened species distribution, see the "Recovering Threatened and Endangered
     Species: Fiscal Years 2005–2006," US Fish & Wildlife Service, https://www.fws.
     gov/endangered/esa-library/pdf/summary__2005-6Recovery.pdf.

6   *Five times in the history of the earth* For more on extinction, check out Elizabeth
     Kolbert, *The Sixth Extinction: An Unnatural History* (New York: Henry Holt, 2014).

9   *"A frontier is a burning edge"* From Gary Snyder, "The Etiquette of Freedom," in *The
     Wilderness Condition: Essays on Environment and Civilization*, ed. Max Oelschlaeger
     (Washington, DC: Island Press, 1992).

10   *the ocean and the tidal marsh are falling* Christopher Craft et al., "Forecasting the
     Effects of Accelerated Sea Level Rise on Intertidal Marshes," *Frontiers in Ecology
     and the Environment* 7, no. 2 (2009), https://doi.org/10.1890/070219.

10   *Here accretion is already being outpaced* Kenneth B. Raposa et al., "Elevation Change
     and the Vulnerability of Rhode Island (USA) Salt Marshes to Sea-Level Rise,"
     Regional Environmental Change 17, no. 2 (February 2017), https://doi.org/10.1007/
     s10113-016-1020-5.

11   *As ice sheets melt, their gravitational pull* The specific impact that each melting ice
     sheet will have on specific locations around the globe is known as the ice sheet's
     "fingerprint." For information on this phenomenon, see recent research that uses
     GRACE satellite data to reveal the unique "fingerprints" of selected ice sheets from
     around the world. In particular: Chia-Wei Hsu and Isabella Velicogna, "Detection
     of Sea Level Fingerprints Derived from GRACE Gravity Data," *Geophysical Research
     Letters* 44, no. 17 (September 2017), https://doi.org/10.1002/2017GL074070.

11 *the places farthest from the largest chunks* Asbury H. Sallenger Jr. et al., "Hotspot of Accelerated Sea-Level Rise on the Atlantic Coast of North America," *Nature Climate Change* 2 (June 2012), https://doi.org/10.1038/nclimate1597.

12 *all the different and conflicting predictions* In 2014, many of the world's leading sea level rise researchers—Robert Kopp, Radley Horton, Christopher Little, Jerry Mitrovica, Michael Oppenheimer, D. J. Rasmussen, Benjamin Strauss, and Claudia Tebaldi—came together to publish an open-access article in *Earth's Future* titled "Probabilistic 21st and 22nd Century Sea-Level Projections at a Global Network of Tide-Gauge Sites," at https://doi.org/10.1002/2014EF000239. Among other things, it offers a good introduction to sea level rise modeling and why different locations will experience different rates of rise.

12 *a series of photo-realistic mock-ups* You too can see these fabulous renderings. There might even be one for a city near you. Check out https://www.choices.climatecentral.org.

13 *Then Ben switches to a rendering* At the time I was writing this, the international community was about to enter into discussions about how to respond to global climate change. The proposals on the table, known as the Paris climate accord, would, it was reported, likely lead to 3.5 degrees Celsius of warming by century's end. While the accord entered into force on November 4, 2016, Donald Trump announced his intention to withdraw the United States from the agreement in August 2017.

13 *I have read about* I read about these things in Elizabeth Kolbert's *The Sixth Extinction*.

13 *the heat waves killing thousands in Paris* Peter Altman, "Killer Summer Heat: Projected Death Roll from Rising Temperatures in America Due to Climate Change," Natural Resources Defense Council, 2012, https://www.nrdc.org/sites/default/files/killer-summer-heat-report.pdf, and Richard C. Keller, *Fatal Isolation: The Devastating Paris Heat Wave of 2003* (Chicago: University of Chicago Press, 2015).

14 *No object thick with pitch* Inspired by Eduardo Cadava, "Trees, Hands, Stars, and Veils: The Portrait in Ruins," in *Fazal Sheikh: Portraits* (Göttingen, Germany: Steidl, 2011).

Persimmons

19 *I walk to the Isle de Jean Charles* The Islanders have done an excellent job collecting primary resources about their community's history. You can check out some of them at https://www.isledejeancharles.com.

19 *Just fifty years ago* Many of the descriptions of Terrebonne Parish in the past are drawn from a month of interviews in and around the Isle de Jean Charles in August 2013 and August 2016 and from Mike Tidwell's *Bayou Farewell: The Rich Life and Tragic Death of Louisiana's Cajun Coast* (New York: Vintage, 2004).

20 *National Oceanographic and Atmospheric Administration had to remap* Nikki Buskey, "As Coast Erodes, Names Wiped Off the Map," *Houma Today*, May 1, 2013, http://www.houmatoday.com/article/DA/20130501/News/608077352/HC/.

23 *In 1951 the first oil rig was installed* Much of the background information on the oil industry and its start on the island comes from photocopied newspaper clippings that Chris Brunet keeps in his home. He is the de facto Island historian. Some of the clippings sourced here are a piece titled "Pirate's Island Rigged for Gold," which appeared in the May 18, 1952, edition of the *New Orleans Times-Picayune*, and "Miles from the End-of-the-Road There's a Community Where Peace, Calm Reign," which also appeared in the *Times-Picayune* though the year is impossible to read.

24 *The oil companies were supposed to "rock" each channel* Ricardo A. Olea and James L. Coleman, "A Synoptic Examination of Causes of Land Loss in Southern Louisiana as Related to the Exploitation of Subsurface Geologic Resources," *Journal of Coastal Research* 30, no. 5 (2014), https://doi.org/10.2112/JCOASTRES-D-13-00046.1.

24 *the channels grow wider, eating into the land* For a comprehensive study on the impact the oil industry's channels have had on Louisiana's wetlands, see Donald F. Boesch et al., "Scientific Assessment of Coastal Wetland Loss, Restoration and Management in Louisiana," *Journal of Coastal Research*, special issue no. 20 (1994), http://www.jstor.org/stable/25735693.

24 *While the dolphin is not direct evidence of sea level rise* It almost is. To learn more, see the NOAA report "Common Bottlenose Dolphin (*Tursiops truncatus truncatus*) Barataria Bay Estuarine System Stock," published in May 2016, at https://www.nefsc.noaa.gov/publications/tm/tm238/319__f2015__bodoBaratariaBay.pdf.

25 *Over the past forty years* This information comes from an in-person interview with the tribal chief, Albert Naquin. He estimated that the population of the Isle de Jean Charles peaked at around 350–400 people in the midseventies, before hurricanes and flooding started forcing people to move in. When I visited in 2013, he estimated that forty full-time residents remained.

25 *put in place to protect Houma* Nikki Buskey, "2010 Will See Unprecedented Levee Spending," *Daily Comet*, December 27, 2009, http://www.dailycomet.com/news/20091227/2010-will-see-unprecedented-levee-spending.

25 *The hundreds of dead cypresses and oaks* For more information on broad-ranging cypress death as a result of saline inundation and storm events throughout coastal Louisiana, see the work of Thomas W. Doyle, the deputy director and ecologist at the United States Geological Survey (USGS) Wetland and Aquatic Research Center in Lafayette, Louisiana.

26 *By 2050 there will be two hundred million* "No Place Like Home: Where Next For Climate Refugees?," Environmental Justice Foundation, 2009, https://ejfoundation.org/reports/no-place-like-home-where-next-for-climate-refugees.

26 *two million of whom* Bob Marshall et al., "Losing Ground," ProPublica, August 2014, http://projects.propublica.org/louisiana/.

28 *According to the United States Geological Survey* Brady R. Couvillion et al., "Land Area Change in Coastal Louisiana from 1932 to 2010," USGS, 2011, https://pubs.usgs.gov/sim/3164/.

28 *lose another 1,750 square miles* Bob Marshall et al., "Louisiana's Moon Shot," ProPublica, December 2014, https://projects.propublica.org/larestoration.

28 *That's because the southern edge* Bob Marshall et al., "Losing Ground."

28 *In wet years a section of the river* For an elegant graphic example of this, check out the map "The Mississippi River Flood of 1927: Showing Flooded Areas and Field Operations," compiled and printed by the US Coast and Geodetic Survey and housed in the National Archives.

29 *the Spanish conquistador Hernando de Soto* Garcilaso de la Vega, *The Florida of the Inca; the Fabulous de Soto Story*, trans. John and Jeannette Varner (Austin: University of Texas Press, 1981). Garcilaso de la Vega was not himself a part of the expedition that he describes.

29 *in 1927 the river inundated an area* Estimates vary for just how much land was inundated and for how long, but most reports suggest that roughly twenty-seven thousand square miles were flooded between April and August of 1927. Christine A. Klein and Sandra B. Zellmer, *Mississippi River Tragedies: A Century of Unnatural Disasters* (New York: New York University Press, 2014).

29 *In an effort to "manage" the mighty river* Jason S. Alexander, Richard C. Wilson, and W. Reed Green, "A Brief History and Summary of the Effects of River Engineering and Dams on the Mississippi River System and Delta," USGS Circular 1375, 2012, https://pubs.usgs.gov/circ/1375/C1375.pdf.

32 *The Chitimacha are said to have lived* Bennett H. Wall, ed., *Louisiana: A History* (Santa Ana, CA: Forum Press, 1990).

32 *In the face of the violence that accompanied* Lewis H. Morgan, "Indian Migrations," *North American Review* 110, no. 226 (January 1870).

32 *But today the high rate of intermarriage* Vernon J. Parenton and Roland J. Pellegrin, "The 'Sabines': A Study of Racial Hybrids in a Louisiana Coastal Parish," *Social Forces* 29, no. 2 (December 1950), https://doi.org/10.2307/2571663.

36 *Others call it the moon bird* Laura Parker, "Arctic Warming Is Shrinking This Adorable Shorebird," *National Geographic*, May 12, 2016, https://news.nationalgeographic.com/2016/05/160512-arctic-warming-shrinking-shorebirds-climate-red-knot/.

36 *Researchers recently found* Jan A. van Gils et al., "Body Shrinkage Due to Arctic Warming Reduces Red Knot Fitness in Tropical Wintering Range," *Science* 352, no. 6287 (May 23, 2016), https://doi.org/10.1126/science.aad6351.

40 *"How to eat"* From Li-Young Lee, "Persimmons," in *Rose* (New York: BOA Editions, 1986).

On Gratitude

This transcription is a composite from two separate in-person interviews with Laura Sewall, the caretaker of the Bates-Morse Mountain Conservation Area and a resident of nearby Small Point. One was conducted on May 11, 2016, at the Coastal Center at Shortridge, and the other was conducted on October 1, 2016, at Laura's house.

The Marsh at the End of the World

General background information for the chapter is from Beverly J. Johnson et al., "The Ecogeomorphology of Two Salt Marshes in Midcoast Maine: Natural History and Human Impacts," Maine Geological Survey, October 2016, http://digitalmaine.com/mgs__publications/26/.

49 *store a quarter of the carbon* Kimbra Cutlip, "For the World's Wetlands, It May Be Sink or Swim. Here's Why It Matters," *Smithsonian*, January 13, 2016, https://www.smithsonianmag.com/smithsonian-institution/worlds-wetlands-it-may-be-sink-or-swim-heres-why-it-matters-180957808/.

49 *That means that an acre of healthy coastal wetlands* Ariana Sutton-Grier, "Coastal Blue Carbon," interview by Troy Kitch, February 12, 2016, in *Making Waves*, NOAA, podcast, https://oceanservice.noaa.gov/podcast/may14/mw124-bluecarbon.html.

50 *a thousand-foot-thick sheet of ice* In New York City, the Wisconsin Ice Sheet was one thousand feet thick; upstate, it was five thousand feet thick. For more information on the geology of the area, check out the New York Parks Department's article "Hot Rocks: A Geological History of New York City Parks," https://www.nycgovparks.org/about/history/geology.

53 *"Some creatures . . . had appeared"* John McPhee coined the phrase "deep time" to describe the Precambrian period in his tome *Annals of the Former World*. The history of geologic discovery included here is a kind of condensation of much of the information found in a fabulous book within the *Annals* called *Basin and Range* (New York: Farrar, Straus and Giroux, 1981).

53 *most nongeologists, me included, are still likely to wildly misidentify* Cinzia Cervato and Robert Frodeman, "The Significance of Geologic Time: Cultural, Educational, and

Economic Frameworks," *Geological Society of America Special Papers* 486 (2012), https://doi.org/10.1130/2012.2486(03).

54 *condensing the history of the planet* Events in the geologic calendar come from "The Geologic Time Scale," University of Kentucky, https://www.uky.edu/KGS/ education/geologictimescale.pdf, and Eleanor Ainscoe, "Every Second Counts," Environmental Research Doctoral Training Fellowship, University of Oxford, https://www.environmental-research.ox.ac.uk/every-second-counts/.   David Brower has long delivered a version of this condensed version of the earth's history in a lecture he playfully titles "The Sermon," likening the history of the planet to the length of Genesis.

54 *twelve-mile-deep sliver* "Biosphere," *National Geographic*, https://www.nationalgeo-graphic.org/encyclopedia/biosphere/.

55 *Global sea levels have risen about nine inches* For a sound scientific introduction to sea level rise, see John A. Church and Neil J. White, "Sea-Level Rise from the Late 19th to the Early 21st Century," *Surveys in Geophysics* 32, no. 4–5 (September 2011), https://doi.org/10.1007/s10712-011-9119-1.

55 *sea levels rose, on average* These figures, which are higher than those provided by the IPCC, are drawn from a lecture by Jerry X. Mitrovica. Mitrovica, the Frank B. Baird, Jr. Professor of Science at Harvard University, argues that the IPCC overestimates the rate of the rise for 1900–1990 and underestimates the rate of the rise since 1990. What Mitrovica's data shows is that this very rate is accelerating faster than we previously thought. See his work in the two following publications: Jerry X. Mitrovica et al., "Probabilistic Reanalysis of Twentieth-Century Global Sea-Level Rise," *Nature* 517 (January 2015), https://doi.org/10.1038/nature14093, and Jerry X. Mitrovica et al., "Reconciling Past Changes in Earth's Rotation with 20th Century Global Sea-Level Rise: Resolving Munk's Enigma," *Science Advances* 1, no. 11 (December 2015), https//doi.org/10.1126/sciadv.1500679.

55 *If you drill into a healthy marsh* Here I am indebted to Evelyn B. Sherr's exhaustive inquiry into coastal wetlands in *Marsh Mud and Mummichogs: An Intimate Natural History of Coastal Georgia* (Athens: University of Georgia Press, 2015).

55 *But when salt water sits* For more on how saline inundation impacts tidal wetlands, look into the research Tiffany Troxler and her students at Florida International University are carrying out in the Everglades, including "Biogeochemical Effects of Simulated Sea Level Rise on Carbon Loss in an Everglades Mangrove Peat Soil," *Hydrobiologia* 726, no. 1 (March 2014), https://doi.org/10.1007/ s10750-013-1764-6.

56 *The US Fish and Wildlife Service didn't understand* Michael Kennish, "Coastal Salt Marsh Systems in the U.S.: A Review of Anthropogenic Impacts," *Journal of Coastal Research* 17, no. 3 (Summer 2001), http://www.jstor.org/stable/4300224.

56   *By the end of the decade following the Depression* Warren S. Bourn and Clarence Cottam, "Some Biological Effects of Ditching Tidewater Marshes," US Fish and Wildlife Service, 1950. By way of Jo Ann Clarke et al., "The Effect of Ditching for Mosquito Control on Salt Marsh Use by Birds in Rowley, Massachusetts," *Journal of Field Ornithology* 55, no. 2 (Spring 1984), http://www.jstor.org/stable/4512881. I should note that ditching was somewhat common beginning in the 1700s, when local farmers used this method to augment the production of salt hay.

56   *The Civilian Conservation Corps didn't care* Ron Rozsa, "Human Impacts on Tidal Wetlands: History and Regulations," in *Tidal Marshes of Long Island Sound: Ecology, History and Restoration*, ed. Glenn D. Dreyer and William A. Niering (New London: Connecticut College Arboretum, 1995).

56   *The standing water in which* Warren S. Bourn and Clarence Cottam, "Some Biological Effects of Ditching Tidewater Marshes," and C. R. Lesser, F. J. Murphy, and R. W. Lake, "Some Effects of Grid System Mosquito Control Ditching on Salt Marsh Biota in Delaware," *Mosquito News* 36, no. 1 (1976).

57   *Cell membranes in the liver* For more on what happens to the human body after death, see Mo Costandi, "Life after Death: The Science of Human Decomposition," *The Guardian*, May 15, 2015, https://www.theguardian.com/science/neurophilosophy/2015/may/05/life-after-death.

58   *"We know that healthy marshes"* Unfortunately I did not have space to include the interesting findings of Cailene Gunn in this chapter. Her thesis shows that when a culverted marsh is restored, it releases significantly less methane than in previous years. For more on this, you should read her honors thesis, "Methane Emissions along a Salinity Gradient of a Restored Salt Marsh in Casco Bay, Maine," at https://scarab.bates.edu/honorstheses/175/.

59   *"How much carbon"* From Kimbra Cutlip, "For the World's Wetlands, It May Be Sink or Swim. Here's Why It Matters."

60   *A molecule of methane* Gayathri Vaidyanathan, "How Bad of a Greenhouse Gas Is Methane?," *Scientific American*, December 22, 2015, https://www.scientificamerican.com/article/how-bad-of-a-greenhouse-gas-is-methane/.

61   *The first reading is 1.55 parts per million* These numbers are taken from Dana Cohen Kaplan's thesis. They were actually gathered on a different day that summer; the numbers from the Science Box were slightly less conclusive the day we went out, which might be related to the water-in-the-lines issue we encountered.

62   *a controversial paper* James Hansen's paper is the coauthored "Ice Melt, Sea Level Rise and Superstorms: Evidence from Paleoclimate Data, Climate Modeling, and Modern Observations That 2°C Global Warming Could Be Dangerous," *Journal of Atmospheric Chemistry and Physics* 16, no. 6 (March 2016), https://doi.org/10.5194/acp-16-3761-2016.

64 *"Out on the islands"* Robert McCloskey, *Time of Wonder* (New York: Puffin Books, 1977).

66 *Because the Gulf of Maine is warmer* Colin Woodard, "Big Changes Are Occurring in One of the Fastest-Warming Spots on Earth," *Portland Press Herald*, October 25, 2015, https://www.pressherald.com/2015/10/25/climate-change-imperils-gulf-maine-people-plants-species-rely/.

67 *When it absorbs carbon dioxide* Victoria J. Fabry et al., "Impacts of Ocean Acidification on Marine Fauna and Ecosystem Processes," *ICES Journal of Marine Science* 65, no. 3 (April 2008), https://doi.org/10.1093/icesjms/fsn048, and Haruko Kurihara, "Effects of CO_2-Driven Ocean Acidification on Early Developmental Phases of Invertebrates," *Marine Ecology Progress Series* 373 (December 2008), https://doi.org/10.3354/meps07802.

Pulse

71 *In 1890, just over six thousand people* This information comes from the 1890 census. The population of south Florida (minus the Keys) was calculated by combining the total population of Dade County (of which Palm Beach was at the time a part) with those of Lee, Desoto, and mainland Monroe Counties. The actual population, according to these calculations, was 7,255. See the census's "Supplement for Florida: Population, Agriculture, Manufacturers, Mines and Quarries," https://www2.census.gov/prod2/decennial/documents/41033935v9-14ch01.pdf.

71 *Since then the wetlands that covered half the state* R. H. Caffey and M. Schexnayder, "Coastal Louisiana and South Florida: A Comparative Wetland Inventory," Interpretive Topic Series on Coastal Wetland Restoration in Louisiana, Coastal Wetland Planning, Protection, and Restoration Act, National Sea Grant Library, 2003, https://lacoast.gov/new/Data/Reports/ITS/Florida/pdf.

71 *the number of black college degree holders* Paul Attewell et al., "The Black Middle-Class: Progress, Prospects, and Puzzles," in *Free at Last?: Black America in the Twenty-First Century*, ed. Juan Battle, Michael Bennett, and Anthony Lemelle (New York: Routledge, 2006).

71 *as did the difference between the average salaries* Before 1900, the majority of businesses were small and run by owners. In the mid-1930s, mandatory reporting of executive pay began. Since 1950, worker pay ratios have increased a thousand-fold. See Barbara Mantel, "Are CEOs Worth the Millions in Compensation They Receive?," *Sage Business Researcher*, July 20, 2015, http://businessresearcher.sagepub.com/sbr-1645-96551-2688702/20150720/executive-pay#. Also see: Marina Gorbis, "To Fix Income Inequality We Need More Than UBI—We Need Universal Basic Assets," *Quartz*, October 11, 2017, https://qz.com/1096659/to-fix-income-inequality-we-need-more-than-ubi-we-need-universal-basic-assets/.

71   *the speed at which we fly* The early flights of the Wright brothers reached an aver-
      age air speed of .013 kilometers per second; the unmanned *Voyager 1* spacecraft,
      launched in 1977, travels 17 kilometers per second. Richard Wagner, *Designs on
      Space: Blueprints for 21ˢᵗ Century Space Exploration* (New York: Simon & Schuster,
      2000).

71   *the combined carbon emissions of the Middle East* "Annex 5A: Trends in International
      Carbon Dioxide Emissions," *Energy—Its Impact on the Environment and Society*,
      National Archives of the United Kingdom, https://webarchive.nationalarchives.
      gov.uk/20060715135302/dti.gov.uk/files/file20356.pdf.

72   *According to Marco Rubio* Nina Burleigh, "Florida Is Sinking. Where Is Marco Rubio?,"
      Mother Jones, February 1, 2016, https://www.motherjones.com/environment/2016/
      02/marco-rubio-climate-change-florida/.

73   *the Intergovernmental Panel on Climate Change* At the heart of all IPCC projections
      are "emission scenarios": low-, mid-, and high-range estimates for future carbon
      emissions. To arrive at the two-feet number, I averaged the rise they predicted
      through their two mid-range models. You can read the chapter on sea level rise:
      John A. Church et al., "Sea Level Change," in *Climate Change 2013: The Physical
      Science Basis: Working Group I Contribution to the Fifth Assessment Report of the
      Intergovernmental Panel on Climate Change* (Cambridge: Cambridge University
      Press, 2013), https://www.ipcc.ch/pdf/assessment-report/ar5/wg1/WG1AR5__
      Chapter13__FINAL.pdf.

73   *The United Nations* Justin Gillis, "Climate Panel Cites Near Certainty on Warming,"
      New York Times, August 19, 2013, http://www.nytimes.com/2013/08/20/science/
      earth/extremely-likely-that-human-activity-is-driving-climate-change-pan-
      el-finds.html.

73   *And the National Oceanic and Atmospheric Administration* Rebecca Lindsey, "Climate
      Change: Global Sea Level," Climate.gov, September 11, 2017, https://www.climate.
      gov/news-features/understanding-climate/climate-change-global-sea-level.

73   *"The rate of sea level rise is currently doubling every seven years"* Just this year, the
      National Academy of Sciences published a paper that shows the rate of sea level
      rise doubling roughly every twenty years: R. S. Nerem et al., "Climate-Change-
      Driven Accelerated Sea Level Rise Detected in the Altimeter Era," *PNAS* (February
      12, 2018), https://doi.org/10.1073/pnas.1717312115. Jerry X. Mitrovica's recent
      work, mentioned in a note above, shows an acceleration on the order of the one Hal
      Wanless describes in his talk, much of which is reproduced here: Harold R. Wanless,
      "The Coming Reality of Sea Level Rise: Too Fast Too Soon," *Sea Level Rise: What's
      Our Next Move?* (Tucson: Institute on Science for Global Policy, 2016), http://sci-
      enceforglobalpolicy.org/wp-content/uploads/56e30928039ff-ISGP%20Sea%20
      Level%20Rise.pdf.

74 *Noah's flood is one* Lydia Barnett, "The Theology of Climate Change: Sin as Agency in the Enlightenment's Anthropocene," *Environmental History* 20, no. 2 (April 2015), https://doi.org/10.1093/envhis/emu131.

74 *Dig into geologic history* Thomas M. Cronin, "Rapid Sea-Level Rise," *Quaternary Science Reviews* 56 (November 2012), https://doi.org/10.1016/j.quascirev.2012.08.021. Also see Jean Liu et al., "Sea-Level Constraints on the Amplitude and Source Distribution of Meltwater Pulse 1A," *Nature Geoscience* 9, no. 2 (February 2016), https://doi.org/10.1038/ngeo2616.

75 *jumping as much as fifty feet* M. E. Weber et al., "Millennial-Scale Variability in Antarctic Ice-Sheet Discharge during the Last Deglaciation," *Nature* 510 (June 5, 2014), https://doi.org/10.1038/nature13397.

75 *"produce earthquakes up to six and seven"* For more information on this phenomenon, see Masaki Kanao et al., "Greenland Ice Sheet Dynamics and Glacial Earthquake Activities," in *Ice Sheets: Dynamics, Formation and Environmental Concerns*, ed. Jonas Müller and Luka Koch (Hauppauge, NY: Nova Science Publishers, 2012), https://www.higp.hawaii.edu/~rhett/PDFs/Kanao__etal__2012__Greenland__Ice__Sheet__Dynamics__and__Glacial__Earthquake__Activities__ICE__SHEETS.pdf.

75 *But a meltwater pulse is the opposite* Vivien Gornitz, *Rising Seas: Past, Present and Future* (New York: Columbia University Press, 2012).

76 *That evening Suzanne Lettieri* Suzanne actually called me a week after I got back from my research trip to Miami. I was at the time driving home from work and I stopped in a parking lot to chat with her. Though I wasn't looking at Miami's skyscrapers, I was certainly thinking about them, and the rabid rate of coastal development taking place despite our knowledge of sea level rise. For more on the condensation of personal narration as part of creative nonfiction, see John D'Agata and Jim Fingal's important book *The Lifespan of a Fact* (New York: W. W. Norton & Company, 2012). Finally, check out Suzanne Lettieri's gorgeous Tumblr site, which shows all different sorts of images of raised homes, at http://post-line.tumblr.com/.

77 *more than 200,000 cubic yards of sand* Jenny Staletovich, "Mayor: Shrinking South Florida Beaches Need Help," *Miami Herald*, July 9, 2015, http://www.miamiherald.com/news/local/environment/article26903749.html.

77 *A century ago* Andres Viglucci, "The 100-Year Story of Miami Beach," *Miami Herald*, March 25, 2015, http://www.miamiherald.com/news/local/community/miami-dade/miami-beach/article15798998.html. For more on the development of Miami Beach, see Wallace Kaufman and Orrin H. Pilkey Jr.'s *The Beaches Are Moving: The Drowning of America's Shoreline*, a prophetic little book published by Duke University Press in 1983.

79 *In 1850 the Swamp Land Act* For more on the Swamp Land Act, see Ann Vileisis, *Discovering the Unknown Landscape: A History of America's Wetlands* (Washington, DC: Island Press, 2012).

80 *But perhaps no place* Again see Ann Vileisis, *Discovering the Unknown Landscape*. Also
    see Stephen S. Light and J. Walter Dineen, "Water Control in the Everglades: A
    Historical Perspective" in *Everglades: The Ecosystem and Its Restoration*, ed. Steven
    M. Davis and John C. Ogden (Boca Raton: St. Lucie Press, 1994). Though the
    sources differ slightly on just how much land was transformed by post-Swamp
    Land Act development, each offers unique insight into how profoundly the legis-
    lation transformed the state.

80 *"Sixteen hundred miles of canals"* From Susan Cerulean and Jono Miller, "The
    Everglades: An Ecology in Five Parts," in *The Book of the Everglades*, ed. Susan
    Cerulean (Minneapolis: Milkweed Editions, 2002).

80 *"swampy, low, excessively hot"* Quoted in Michael Grunnwald, "A Requiem for Florida,
    the Paradise That Should Never Have Been," *Politico*, September 8, 2017, https://
    www.politico.com/magazine/story/2017/09/08/hurricane-irma-florida-215586.

80 *multiple billion-dollar industries* For more on Florida's citrus production, see the
    findings of the Florida Department of Citrus and the University of Florida:
    Christa D. Court et al., "Economic Contributions of the Florida Citrus Industry
    in 2015–16," May 9, 2017, http://fred.ifas.ufl.edu/pdf/economic-impact-analysis/
    Economic__Impacts__of__the__Florida__Citrus__Industry__2015__16.pdf.
    Governor Rick Scott applauded Florida's tourism industry as justification for
    signing a new timeshare law into being on his website: https://www.flgov.com/
    governor-scott-applauds-floridas-tourism-marketing-2/. Finally, Broward County
    points to seniors fueling the local economy on their website: http://www.adrcbro-
    ward.org/economicimpact.php.

84 *high-rises currently under construction* Please read Nathan Brooker's excel-
    lent analysis of sea level rise's potential impact on the property market in
    south Florida, titled: "Miami Beach: Property Market Braced for Change
    in Climate," *Financial Times*, January 4, 2017, https://www.ft.com/content/
    de73b604-c12b-11e6-81c2-f57d90f6741a.

84 *I imagine all of it underwater* For a highly specific and important study on the in-
    crease in flooding events on Miami Beach since 2000, see Shimon Wdowinski
    et al., "Increasing Flooding Hazard in Coastal Communities Due to Rising Sea
    Level: Case Study of Miami Beach, Florida," *Ocean & Coastal Management* 126
    (June 2016), https://doi.org/10.1016/j.ocecoaman.2016.03.002.

86 *All along the East Coast* Jonathan Corum recently authored an interactive piece on
    the phenomenon: "A Sharp Increase in 'Sunny Day' Flooding," *New York Times*,
    September 3, 2016, https://www.nytimes.com/interactive/2016/09/04/science/
    global-warming-increases-nuisance-flooding.html.

87 *Like Miami Beach, Shorecrest was built* Robert A. Renken et al., "Impact of
    Anthropogenic Development on Coastal Ground-Water Hydrology in Southeastern
    Florida, 1900–2000," USGS, 2005, https://pubs.usgs.gov/circ/2005/circ1275/.

87 *On the strip, where billions of dollars* For more on the city's $500 million effort to stay dry, check out the interactive web feature "Pump It," by Fusion Media Group, http://interactive.fusion.net/pumpit/. For something more technical, I recommend the City of Miami Beach's "Stormwater Management Master Plan," created by CDM Smith and published in 2012, at http://www.miamibeachfl.gov/city-hall/city-manager/master-plans/. The *Miami Herald* estimates that property values in Miami Beach equaled over $5 billion in 2015. See Joey Flechas, "Property Values Surge in Miami Beach with $1.1 Billion in New Construction," *Miami Herald*, June 3, 2016, http://www.miamiherald.com/news/local/community/miami-dade/miami-beach/article81419997.html.

87 *But in Shorecrest, Hialeah, and Sweetwater* You can find many of the maps the Home Owners' Loan Corporation used for redlining, including their map of Miami, in an archive called "Mapping Inequality," hosted by the University of Richmond at https://dsl.richmond.edu/panorama/redlining/#loc=4/36.71/-96.93&opacity=0.8&text=intro.

87 *the discriminatory banking practice known as redlining* For more information on the impact of redlining on communities of color, see Ta-Nehisi Coates's revelatory article "The Case for Reparations," *The Atlantic*, June 2014, https://www.theatlantic.com/magazine/archive/2014/06/the-case-for-reparations/361631/.

88 *"It is the destruction of the world"* This text is actually the first three lines of "II: It is the destruction of the world," the second "sabbath poem" Wendell Berry wrote in 1988. You can find it in *This Day: Collected and New Sabbath Poems* (Berkeley: Counterpoint, 2013).

91 *The last time carbon dioxide levels were this high* "The Pliocene Epoch," University of California Museum of Paleontology, http://www.ucmp.berkeley.edu/tertiary/pliocene.php.

91 *Shifting plate tectonics reopened the Strait of Gibraltar* Katherine Kornei, "A Megaflood-Powered Mile-High Waterfall Refilled the Mediterranean," *Scientific American*, March 26, 2018, https://www.scientificamerican.com/article/a-megaflood-powered-mile-high-waterfall-refilled-the-mediterranean-video/.

91 *But no one can remember these things* Andrew Freedman, "The Last Time CO_2 Was This High, Humans Didn't Exist," Climate Central, May 3, 2013, http://www.climatecentral.org/news/the-last-time-co2-was-this-high-humans-didnt-exist-15938.

91 *Fifteen thousand years ago, human beings were transforming* It is difficult to date the *exact* start of agriculture. Often we try to link it to the rise of sedentism. Some sources suggest that protofarming could have started as many as twenty-three thousand years ago: Ainit Snir et al., "The Origin of Cultivation and Proto-Weeds, Long before Neolithic Farming," *PLoS One* (July 22, 2015), https://doi.org/10.1371/journal.pone.0131422. Much more widely accepted is a start date around twelve thousand years ago, in the Fertile Crescent. For more on the link between the rise of agriculture and meltwater pulses, see James C.

Scott's *Against the Grain: A Deep History of the Earliest States* (New Haven: Yale University Press, 2017).

91 *Fifteen thousand years ago, we domesticated the first pig* This number is an approximation. The following article suggests that the earliest domestication of this species took place *more* than twelve centuries ago: Jean-Denis Vigne et al., "Pre-Neolithic Wild Boar Management and Introduction to Cyprus More Than 11,400 Years Ago", *PNAS* 106, no. 38 (September 2009), https://doi.org/10.1073/pnas.0905015106.

91 *the woolly mammoth started to go locally extinct* Often a species will go "locally extinct" before global extinction. Hundreds of large mammal species disappeared during the transition from the last glaciation into the Holocene, from around thirty thousand to five thousand years ago. This is often referred to as the "Quaternary Extinction Event." Research links these extinction events to the concurrence of dramatic environmental change (including Meltwater Pulse 1A) and the spread of humans worldwide. For an excellent overview of the topic, see Paul L. Koch and Anthony D. Barnosky, "Late Quaternary Extinctions: State of the Debate," *Annual Review of Ecology, Evolution, and Systematics* 37 (2006), https://doi.org/10.1146/annurev.ecolsys.34.011802.132415. If you are interested in the specifics of local extinction, check out S. David Webb, ed., *The First Floridians and Last Mastodons: The Page-Ladson Site in the Aucilla River* (Dordrecht, Netherlands: Springer, 2006).

92 *"What do you expect me to do"* From Katie Ford, "Flee," in *Colosseum* (Minneapolis: Graywolf Press, 2008).

On Reckoning

Transcribed from a telephone interview with Dan Kipnis on March 4, 2016. Dan is the chairman of the Miami Beach Marine and Waterfront Protection Authority. I spoke with him again in late 2017; at that time, he still hadn't sold his home, and noted that many other homes on his block were then for sale, flooding the market and making it more difficult to move. He has lowered the asking price multiple times, and fears that he will no longer be able to sell before the coastal housing market crashes.

95 *60 percent of those households* Dan Kipnis does not exaggerate. See "Income & Poverty in Miami-Dade County: 2013," Department of Regulatory & Economic Resources, Miami-Dade County, June 2015, https://www.miamidade.gov/business/library/reports/2013-income-poverty.pdf.

On Storms

Transcribed from an interview with Nicole Montalto (now Buonamano) on October 30, 2014. I wrote the article that Nicole references, "Fox Beach Fades to Green: After Superstorm Sandy, Staten Island Neighbors Give Their Homes to Nature," at *Al Jazeera America*. A photo of her home was the feature image: http://projects.aljazeera.com/2014/fox-beach/.

113 *Both the size of the storm and its unusual route* For more on how unprecedented Sandy was, read Adam Sobel's *Storm Surge: Hurricane Sandy, Our Changing Climate, and Extreme Weather of the Past and Future* (New York: Harper Wave, 2014).

115 *over a hundred million dollars* Funding for buyouts comes from multiple sources. State and local governments regularly use federally allocated recovery funds (often in the form of Community Development Block Grants) to provide the 25 percent match required by FEMA to participate in the Hazard Mitigation Grant Program. For an in-depth overview of the history of this recovery strategy, see Robert Freudenberg et al., "Buy-In for Buyouts: The Case for Managed Retreat from Flood Zones," Lincoln Institute of Land Policy, 2016, https://www.lincolninst.edu/sites/default/files/pubfiles/buy-in-for-buyouts-full.pdf. More specifically, in Staten Island funding was initially made available by Governor Cuomo to purchase and demolish homes impacted by Sandy in February 2013: Thomas Kaplan, "Cuomo Seeking Home Buyouts in Flood Zones," *New York Times*, February 3, 2013, http://www.nytimes.com/2013/02/04/nyregion/cuomo-seeking-home-buyouts-in-flood-zones.html. The original "buyout zone" in Oakwood Beach would eventually be expanded to include Ocean Breeze and Graham Beach.

116 *Twenty-two thousand years ago* "Chapter Fifteen: East and South Shores of Staten Island," in *A Stronger, More Resilient New York*, New York City Special Initiative for Rebuilding and Resiliency, June 2013, http://www.nyc.gov/html/sirr/html/report/report.shtml. Also see Makan A. Karegar, Timothy H. Dixon, and Simon E. Engelhart, "Subsidence along the Atlantic Coast of North America: Insights from GPS and Late Holocene Relative Sea Level Data," *Geophysical Research Letters* 43, no. 7 (April 2016), https://doi.org/10.1002/2016GL068015.

116 *At the turn of the last century* Most of the ecological backstory of New York City is drawn from Ted Steinberg's *Gotham Unbound: The Ecological History of Greater New York* (New York: Simon & Schuster, 2015).

116 *roughly 90 percent of the city's wetlands* Plan NYC, "New York City's Wetlands Strategy," Mayor's Office of Long-Term Planning and Sustainability, May 2012, http://www.nyc.gov/html/planyc2030/downloads/pdf/nyc__wetlands__strategy.pdf.

116 *Chinatown was once a wetland* For more on the preindustrial ecology of the region, see Eric W. Sanderson's astonishing book *Mannahatta: A Natural History of New York City* (New York: Abrams, 2009). The Regional Planning Association has created a wetland-specific interactive map showing wetland loss in New York City, "The Region's Coastal Wetlands: Past to Present." You can access it here: https://rpany.maps.arcgis.com/apps/StorytellingSwipe/index.html?appid=bf6bdccb1a-f345ad99c4d01b9c2629ac. For more on Staten Island's wetlands, see note for page 127.

117 *It's not just Gotham* Thomas E. Dahl and Gregory J. Allord, "History of Wetlands in the Conterminous United States," *National Water Summary on Wetland Resources,*

United States Geological Survey, March 1997, https://water.usgs.gov/nwsum/WSP2425/history.html.

117 *a significant portion of these wetlands turned landfills* This information comes from two separate conversations, one with Ted Steinberg on September 18, 2014, the other with Eric Klinenberg, director of the Institute for Public Knowledge at New York University, on September 16, 2014.

120 *To most, a wetland is just a mess of grass* Michael Taussig's *My Cocaine Museum* (Chicago: University of Chicago Press, 2004) includes a couple of chapters that provide a cultural history of swamps.

124 *Typically HMGP grants are used to relocate rural riverine communities* Liz Koslov's excellent work on retreat has informed this chapter. Please read her article "The Case for Retreat," *Public Culture* 28, no. 2 (2016), https://doi.org/10.12.15/08992363-3427487. Also worthy of a spirited mention is Liz's forthcoming book, *Retreat: Moving to Higher Ground in a Climate-Changed City*, from University of Chicago Press.

126 *The developers received variances* John Rudolf et al., "Hurricane Sandy Damage Amplified by Breakneck Development of Coast," *Huffington Post*, November 2012, https://www.huffingtonpost.com/2012/11/12/hurricane-sandy-damage__n__2114525.html.

126 *exempting them from a 1973 law* The law is New York State's "Tidal Wetland Act," which builds on the federal "Clean Water Act" enacted in 1972. For more on both, take a look at "New York City's Wetlands: Regulatory Gaps and Other Threats," Plan NYC, 2009, http://www.nyc.gov/html/om/pdf/2009/pr050-09.pdf.

127 *Staten Island's eastern shore* This information comes from my conversation with Alan Benimoff and is largely supported by this spectacular 1888 map, produced by the USGS, that illustrates the island's extensive wetlands: "Staten Island, Survey of 1888–89 and 1897, Ed. of 1900, Repr. 1908," https://digitalcollections.nypl.org/items/ba0dabff-e7b5-ecfc-e040-e00a18061295.

127 *Oakwood Beach and Ocean Breeze both sit* Matthew Schuerman, "Deadly Topography: The Staten Island Neighborhood Where Eleven Died during Sandy," WNYC, February 25, 2013, https://www.wnyc.org/story/271288-tricked-topography-how-staten-island-neighborhood-became-so-dangerous-during-sandy/.

131 *another tremendously vulnerable tidal wetland* Matthew L. Kirwan, "Tidal Wetland Stability in the Face of Human Impacts and Sea-Level Rise," *Nature* 504, no. 7478 (December 5, 2013), https://doi.org/10.1038/nature12856.

131 *The ancients knew that* Umberto Quattrocchi, "Spartina Shreber," *CRC World Dictionary of Grasses: Common Names, Scientific Names, Eponyms, Synonyms and Entomology* (Boca Raton: Taylor & Francis, 2006).

131 *A knotted rope, to measure* Cristian Violatti, "Greek Mathematics," *Ancient History Encyclopedia*, https://www.ancient.eu/article/606/greek-mathematics/.

131 *A rope painted red to mark* "The Red Rope," British Museum, http://www.ancient-greece.co.uk/athens/challenge/cha__set.html.

131 *A twisted rope to launch* Hans Michael Schellenberg, "Catapult," in *Conflict in Ancient Greece and Rome: The Definitive Political, Social, and Military Encyclopedia*, ed. Sara E. Phang et al. (Santa Barbara: ABC-CLIO, 2016).

131 *This is the derivation of its genus name* Here, again, I am indebted to Evelyn B. Sherr's *Marsh Mud and Mummichogs*.

132 *The sediment around them loosens* For more on the ongoing research linking sea level rise with peat collapse, check out the work that Tiffany Troxler is doing at Florida International University. In particular, see Lisa G. Chambers, Stephen Davis, and Tiffany G. Troxler, "Sea Level Rise in the Everglades: Plant-Soil-Microbial Feedbacks in Response to Changing Physical Conditions," in *Microbiology of the Everglades Ecosystem*, ed. James A. Entry et al. (Boca Raton: CRC Press, 2015).

132 *What distinguishes rhizomes* Cheol Seong Jang et al., "Functional Classification, Genomic Organization, Putatively Cis-Acting Regulatory Elements, and Relationship to Quantitative Trait Loci, of Sorghum Genes with Rhizome-Enriched Expression," *Plant Physiology* 142, no. 3 (November 2006), https://doi.org/10.1104/pp.106.082891.

On Vulnerability

Transcribed from two interviews (the first in person and the second on the phone) with Marilynn Wiggins, on August 20, 2016, and on January 25, 2017.

133 *Plus on top of it all* For information on the wastewater treatment plant and the mosquito control plant's impact on the community, check out Jim Little, "Opening Delayed for Pensacola's Corinne Jones Park," *Pensacola News Journal*, May 10, 2017, http://www.pnj.com/story/news/2017/05/10/opening-delayed-pensacolas-corin ne-jones-park/101483940/.

Risk

137 *by the time I reach Alvin Turner's double-wide* I have changed the names of every single person who appears in this chapter, for different reasons: of Alvin Turner and Zoey, to protect the identities of those who spoke to me in confidence; of Robert Brown, to protect the identity of someone I could not find; and of Samuel, at my publisher's recommendation, to avoid potential liability.

137 *one of the city's lowest-lying neighborhoods* The historical information in this section is drawn from two sources: "A History of the Pensacola Police Department," compiled by Sergeant Michael Simmons, available at http://www.cityofpensacola.

com/947/History-of-the-Pensacola-Police-Departme, and Scott Satterwhite, "On the Pensacola Waterfront," *Pensacola Independent News*, August 2, 2007, http://in-weekly.net/article.asp?artID=5023.

137 *Long before the Tanyard* Russell Burdge and Brandon Tidwell, "Bruce Beach (Pensacola, FL) Restoring Community and Habitat along the Historic City of Pensacola's Bayfront" (poster, Restore America's Estuaries National Conference and Expo, Tampa, FL, October 20–24, 2012). Also see Joseph Purcell, cartographer, *A Plan of Pensacola and Its Environs in Its Present State, from an Actual Survey in 1778*, 1778, 51 x 72 cm, Library of Congress, https://www.loc.gov/item/73691620/. And finally, for information on the company (Panton, Leslie & Co.) that ran the tanyard, see Lawrence S. Rowland, Alexander Moore, and George C. Rogers Jr., *The History of Beaufort County, South Carolina: Vol. 1, 1514–1861* (Columbia: University of South Carolina Press, 1996).

137 *When Hurricane Ivan spun* Felicity Barringer and Andrew C. Revkin, "HURRICANE IVAN: THE OVERVIEW; Hurricane's Fury Kills 23 Along Gulf," *New York Times*, September 17, 2004, https://query.nytimes.com/gst/fullpage.html?res=9E0D-E3D61639F934A2575AC0A9629C8B63&sec=&spon=&pagewanted=1.

139 *Wetlands have long been viewed* Rod Giblett, *Postmodern Wetlands: Culture, History, Ecology* (Edinburgh: Edinburgh University Press, 1996). Also see Lawrence N. Powell, *The Accidental City: Improvising New Orleans* (Cambridge: Harvard University Press, 2012).

139 *"The exhalations that continually rise"* William Byrd, quoted in Rod Giblett, *Postmodern Wetlands*.

140 *The insurance industry relies* I am deeply indebted to the work of Rebecca Elliott in this chapter—both in the many phone and email conversations we have shared and in her phenomenal paper "Who Pays for the Next Wave? The American Welfare State and Responsibility for Flood Risk," *Politics & Society* 45, no. 3 (September 2017), https://doi.org/10.1177/0032329217714785.

140 *The NFIP was founded in 1968* Stuart Mathewson et al., "The National Flood Insurance Program: Past, Present . . . and Future?," American Academy of Actuaries, 2011, https://www.actuary.org/pdf/casualty/AcademyFloodInsurance__Monograph__110715.pdf.

140 *increased by a factor of four* This number uses an estimate of homes in the flood-plain from 1960—pre-dating the start of the NFIP by eight years—as its point of comparison. Caroline Peri, Stephanie Rosoff, and Jessica Yager, "Population in the U.S. Floodplains," NYU Furman Center, December 2017, https://furmancenter.org/files/Floodplain__PopulationBrief__12DEC2017.pdf.

140 *Today roughly fifteen million homes* Caroline Peri, Stephanie Rosoff, and Jessica Yager, "Population in the U.S. Floodplains." Also of note: the percentage of the population living at or below the poverty level is the same nationwide as it is in the floodplain, 15 percent.

140 *"special hazard flood areas"* As the likelihood of storm events is on the rise, many are pushing for a move away from using the term "hundred-year floodplain" because, they argue, it creates a false sense of security. Today those located within the "hundred-year floodplain" (now often referred to as "special hazard flood areas") are statistically expected to flood closer to once every seventy years, with the interval between events expected to grow even smaller in the near future. Or, put another way, if you have a thirty-year mortgage, your property has a 26 percent chance of flooding during the lifetime of your loan.

140 *On the East Coast alone* Brooke Jarvis, "When Rising Seas Transform Risk Into Certainty," *New York Times*, April 18, 2017, https://www.nytimes.com/2017/04/18/magazine/when-rising-seas-transform-risk-into-certainty.html.

140 *Sorry for digging* This section is a riff on Jamaal May's always-gutting poem "Ode to the White-Line-Swallowing Horizon," in *The Big Book of Exit Strategies* (Farmington, ME: Alice James Books, 2016).

141 *"Empathy is always perched"* All quotes here and later from Leslie Jamison, "The Empathy Exams," in *The Empathy Exams* (Minneapolis: Graywolf Press, 2014).

142 *The tiny granules* Ginger M. Allen and Martin B. Main, "Florida's Geological History," Publication WEC189, Wildlife Ecology and Conservation Department, University of Florida, April 2005, http://edis.ifas.ufl.edu/uw208.

142 *"will never wholly kiss you"* This line and the others tattooed on my back come from the poem "since feeling is first" in E. E. Cummings's third collection, *is 5* (New York: Liveright, 1926).

143 *The NFIP was already $24 billion in debt* For more on this topic, once again please look at Rebecca Elliott's work, as well as Rachel Cleetus, "Overwhelming Risk: Rethinking Flood Insurance in a World of Rising Seas," Union of Concerned Scientists, August 2013, https://www.ucsus.org/global__warming/science__and__impacts/impacts/flood-insurance-sea-level-rise.html#.WocogJO7-qQ.

143 *Then Hurricane Harvey dumped* The following articles on Harvey shaped this section: Jason Samenow and Matthew Cappucci, "Rains from Harvey Obliterate Records, Flood Disaster to Expand," *Washington Post*, August 28, 2017, https://www.washingtonpost.com/news/capital-weather-gang/wp/2017/08/28/rains-from-harvey-obliterate-records-flood-disaster-to-expand/?utm__term=.3f219934e70f; and "FEMA Expects More Than 450,000 Harvey Disaster Victims to File for Assistance," CNBC, August 28, 2017, https://www.cnbc.com/2017/08/28/fema-expects-more-than-450000-harvey-disaster-victims-to-file-for-assistance.html.

143 *over one trillion gallons of water* This estimate is from the Harris County Flood Control District (https://www.hcfcd.org/hurricane-harvey/). The total rainfall from Harvey across the continental United States is closer to thirty-three trillion gallons, as estimated by the *Washington Post*. The newspaper also created an interesting graphic to illustrate just what thirty-three trillion gallons of water looks like. You can see it here:

Angela Fritz and Jason Samenow, "Harvey Unloaded 33 Trillion Gallons of Water in the U.S.," *Washington Post*, September 2, 2017, https://www.washingtonpost.com/news/capital-weather-gang/wp/2017/08/30/harvey-has-unloaded-24-5-trillion-gallons-of-water-on-texas-and-louisiana/?utm_term=.5dd918b8de27.

144 *private insurers will choose* Michael Thrasher, "The Private Flood Insurance Market Is Stirring after More Than 50 Years of Dormancy," *Forbes*, August 26, 2016, https://www.forbes.com/sites/michaelthrasher/2016/08/26/the-private-flood-insurance-market-is-stirring-after-more-than-50-years-of-dormancy/#4c2794736dda.

146 *But Pensacola never developed* Matthew J. Clavin, *Aiming for Pensacola: Fugitive Slaves on the Atlantic and Southern Frontiers* (Cambridge: Harvard University Press, 2015).

146 *Pensacola and its surrounding area were as close* Jane Landers, *Black Society in Spanish Florida* (Champaign: University of Illinois Press, 1999).

146 *By the end of the nineteenth century* "Preservation District Guidelines & Recommendations: Pensacola, Florida," University of West Florida Historic Trust, 2014, http://www.cityofpensacola.com/AgendaCenter/ViewFile/Agenda/07062015-936.

148 *"with thirty inches of rain falling in twenty-four hours"* During August 2016, epic rains inundated the Gulf Coast. This information comes from notes of my conversation with Alvin in his living room, though it appears that the estimates of the amount of rain that fell over a twenty-four-hour period of time were slightly exaggerated: 31.39 inches of rain fell on Watson, Louisiana—twenty miles north of Baton Rouge—but over a period of days. Tom Di Liberto, "August 2016 Extreme Rain and Floods along the Gulf Coast," Climate.gov, August 19, 2016, https://www.climate.gov/news-features/event-tracker/august-2016-extreme-rain-and-floods-along-gulf-coast.

148 *as much as two feet of rain fell* Bill Cotterell, "U.S. Gulf Coast Hit by Flooding after Twenty-Four Hours Non-stop Rain," *Reuters*, April 29, 2014, https://www.reuters.com/article/us-usa-tornado-weather/u-s-gulf-coast-hit-by-flooding-after-24-hours-non-stop-rain-idUSBREA3R08320140430.

149 *Risk is a word* Molly Wallace's *Risk Criticism: Precautionary Reading in an Age of Environmental Uncertainty* (Ann Arbor: University of Michigan Press, 2016) is an insightful introduction to how different kinds of risk have shaped our literary representations of the world since the beginning of the twentieth century.

149 *"possibility of injury"* Merriam-Webster, s.v. "risk," accessed February 16, 2018, https://www.merriam-webster.com/dictionary/risk.

150 *"unfortunate event[s]"* From Rebecca Elliott, "Who Pays for the Next Wave? The American Welfare State and Responsibility for Flood Risk."

150 *eight out of the ten most expensive hurricanes* Skye Gould and Jonathan Garber, "Mapped: The 10 Costliest Hurricanes in US History," *Business Insider*, September

8, 2017, http://www.businessinsider.com/hurricane-irma-costliest-hurricanes-us-history-map-2017-9.

153 *"One of the cruelest things"* Jia Tolentino, "How Men Like Harvey Weinstein Implicate Their Victims in Their Acts," *New Yorker*, October 11, 2017, https://www.newyorker.com/culture/jia-tolentino/how-men-like-harvey-weinstein-implicate-their-victims-in-their-acts. The framing of the second half of this paragraph is modeled on Tolentino's writing as well.

154 *"can do anything"* To read the entire conversation he had with Billy Bush of *Access Hollywood*, read "Transcript: Donald Trump's Taped Comments About Women," *New York Times*, October 8, 2016, https://www.nytimes.com/2016/10/08/us/donald-trump-tape-transcript.html.

155 *In epidemiology, belonging to a "risk group"* Susan Sontag, *AIDS and Its Metaphors* (New York: Farrar, Straus and Giroux, 1989).

156 *"That victim who is able to articulate"* James Baldwin, *The Devil Finds Work* (New York: Vintage, 1976).

156 *Written by those who earn* Eileen Patten, "Racial, Gender Wage Gaps Persist in the U.S. despite Some Progress," Pew Research Center, July 1, 2016, http://www.pewresearch.org/fact-tank/2016/07/01/racial-gender-wage-gaps-persist-in-u-s-despite-some-progress/.

159 *an insurance policy they cannot afford* Marilyn Montgomery, "Affordability of Flood Insurance in Pensacola, Florida," prepared for the Florida Department of Emergency Management, Escambia County, and City of Pensacola. This study shows that 30 percent of the homes in the AE zone—a designation that signifies the home is likely to flood but will not be significantly impacted by wave action—as the Tanyard largely is, cannot afford the risk-based flood insurance premiums offered by the NFIP. Additionally, if the homes rest in an area where additional surge protection is needed, the percentage of households that cannot afford the premium jumps to 58 percent.

159 *In Gulf Breeze I met a couple* This information comes from an interview with a couple who lives on Coral Strip Parkway and had recently undergone an extensive home-lifting project. The interview was conducted in person and over the phone on two separate dates: August 23 and September 14, 2016.

161 *but didn't I flee* There are echoes of Katie Ford's poem "Flee" in this sentence. Please read *Colosseum* for a lyrically minded account of Hurricane Katrina and New Orleans during and after the storm.

161 *I am done dreaming* With thanks to Jane Costlow, who wisely, after reading this chapter, reminded me that ours is the era of security's end.

161 *"It's a terrible delusion to think"* James Baldwin, "Words of a Native Son," in *Collected Essays* (New York: Library of America, 1998).

On Opportunity

Transcribed from an in-person interview with Chris Brunet on the Isle de Jean Charles on August 23, 2016.

Goodbye Cloud Reflections in the Bay

167 *For a long time, I thought the story* This chapter is the result of a second research trip I made to the Island in August 2016, funded in part by the Bates College Faculty Development Fund.

170 *Then came the National Disaster Resilience Competition* For more information on the competition, see the HUD Exchange website, and in particular its "National Disaster Resilience" entry: https://www.hudexchange.info/programs/cdbg-dr/resilient-recovery/.

170 *likely even those who had left* This detail about who, precisely, will be invited to participate in the relocation remains hotly contested.

177 *Months later I am asked to respond* This lecture was hosted by the Happold Foundation and held at the New School on April 5, 2017, and was titled "Rebuilding or Relocating: How to Respond to Climate Change."

Connecting the Dots

185 *The rufous hummingbird is no larger* Facts about the rufous are sourced from a smattering of websites about hummingbirds, including the Audubon Society (http://www.audubon.org/field-guide/bird/rufous-hummingbird), the US Fish and Wildlife Service (https://www.fws.gov/pollinators/Features/Rufous.html), and the Arizona-Sonora Desert Museum (https://www.desertmuseum.org/pollination/).

185 *travel five thousand miles each year* This number is calculated by taking the round-trip distance between the Isle de Jean Charles and the H. J. Andrews Experimental Forest (5,028 miles) on bicycle, since the rufous's route isn't likely to be direct. Some rufous travel nearly four thousand miles *one way* between their southernmost wintering grounds in Mexico and their northernmost breeding grounds in Alaska. See "Rufous Hummingbird and Spring Migration," Journey North, https://www.learner.org/jnorth/tm/humm/sl/17/article.html.

185 *the Gulf Coast* Increasingly the rufous is being spotted in the American South as far east as Florida. This was uncommon fifty years ago, and it is unclear whether they always wintered in this area or if it represents a shift in their winter range. See the Cornell Lab of Ornithology for more information, in particular "Rufous Hummingbird," by William Calder, in their *Birds of North America* feature, available online: https://birdsna.org/Species-Account/bna/species/rufhum/introduction. Also check out this detailed study on the shift in the birds' wintering and migratory routes: Geoffrey E. Hill, Robert R. Sargent, and Martha B. Sargent, "Recent

Change in the Winter Distribution of Rufous Hummingbirds," *The Auk* 115, no. 1 (January 1998), https://doi.org/10.2307/4089135.

186 *8,849 spools a year* This number is calculated by taking the round-trip distance between the Isle de Jean Charles and the H. J. Andrews Experimental Forest (5,028 miles, as estimated above), converting it to feet (26,547,840), and then dividing by a thousand-yard KINGSO spool of thread (3,000 feet.)

187 *Long-Term Ecological Reflections program* Please read the grounded compilation of essays that traces the first decade of the program, *Forest Under Story: Creative Inquiry in an Old-Growth Forest*, ed. Nathaniel Brodie, Charles Goodrich, and Frederick J. Swanson (Seattle: University of Washington Press, 2016).

189 *a new three-part long-term study* Much of the information on this study is compiled from time in the field with Sarah Frey and her researchers, as well her article "Microclimate Predicts Within-Season Distribution Dynamics of Montane Forest Birds," *Diversity and Distributions* 22, no. 9 (September 2016), https://doi.org/10.1111/ddi.12456.

190 *According to the Audubon Society* "Climate Endangered: Rufous Hummingbird," *The Climate Report*, Audubon Society, http://climate.audubon.org/birds/rufhum/rufous-hummingbird. Many rufous also winter in the equatorial tropics, and in the scrubland of Mexico in particular, though this study is limited to the United States. Also see Andrew Lawler, "Rufous Hummingbirds Turning Up in Unusual Places," *Audubon* (March–April 2010), http://www.audubon.org/magazine/march-april-2010/rufous-hummingbirds-turning-unusual-places.

190 *the bodies of these small birds do the connecting* William Calder, "Rufous and Broad-tailed Hummingbirds: Pollination, Migration, and Population Biology," in *Conserving Migratory Pollinators and Nectar Corridors in Western North America*, ed. Gary Paul Nabhan (Tucson: University of Arizona Press, 2004).

192 *the federal ornithologist Frederick Lincoln* Robert M. Wilson, *Seeking Refuge: Birds and Landscapes of the Pacific Flyway* (Seattle: University of Washington Press, 2010).

192 *In North America, a third of these nomadic creatures* "State of North America's Birds 2016," North American Bird Conservation Initiative, 2016, http://www.stateofthebirds.org/2016/.

192 *the chicken-or-the-egg question in reverse* The numbers of rufous have been dropping steadily. According to the North American Breeding Bird Survey, the number of rufous have declined by 62 percent since 1966. That being said, the rufous is not on the endangered species list.

193 *Species extinction, of course, means not individual deaths* The thinking in this paragraph is very much indebted to Srinivas Aravamudan's powerful essay "The Catachronism of Climate Change" in the special climate change edition of *diacritics* (2013), https://doi.org/10.1353/dia.2013.0019.

195 *The shrub is so prickly* This information, and almost all of the other information on plants of the Pacific Northwest in this chapter, comes from Jim Pojar and Andy MacKinnon's aptly named and wonderfully navigable *Plants of Coastal British Columbia including Washington, Oregon & Alaska* (Vancouver: Lone Pine Publishing, 2005). It's a cult classic in the region.

196 *We are snakes with our own tails* This idea is explored in Annie Dillard's transcendent, Pulitzer Prize–winning book, *Pilgrim at Tinker Creek* (New York: Harper Perennial, 1974).

197 *Sea levels were 550 feet higher* Estimates of the height of sea levels during the Late Cretaceous period vary from about 131 feet to 820 feet above the present level. I decided to use the average that R. Dietmar Müller et al. estimate in "Long-Term Sea-Level Fluctuations Driven by Ocean Basin Dynamics," *Science* 319, no. 5868 (March 2008), http://www.jstor.org/stable/20053529.

197 *but also one-third of the planet's dry land* This rough estimation comes from *Encyclopedia Britannica Online*, s.v. "Cretaceous Period," accessed March 21, 2018, https://www.britannica.com/science/Cretaceous-Period.

198 *"Doing science with awe"* The idea of scientific observation as reciprocity is explored in much greater length in Robin Wall Kimmerer's wise and much-needed book *Braiding Sweetgrass: Indigenous Wisdom, Scientific Knowledge, and the Teachings of Plants* (Minneapolis: Milkweed Editions, 2013).

199 *It is a labor-intensive task* This musing on the act of "paying attention" has been informed by many talks with Laura Sewall.

199 *"the great conversation"* This idea is explored further in Thomas Berry's book (cowritten with Thomas Clarke), *Befriending the Earth: A Theology of Reconciliation between Humans and the Earth* (Mystic, CT: Twenty-Third Publications, 1991).

200 *two thousand pairs of spotted owls left* The statistic comes from "Basic Facts about Northern Spotted Owls," Defenders of Wildlife, https://defenders.org/northern-spotted-owl/basic-facts. The number of known breeding pairs in the Andrews comes from a conversation with Steve Ackers, the head of the northern spotted owl research team based at the site.

200 *During the "Forest Wars"* All information on the history of the Andrews is sourced from a talk with Frederick H. Swanson, professor of forest ecosystems and society at the University of Oregon, on May 30, 2016, and from *Forest Under Story*.

201 *an injunction against harvesting* It should be noted that the lumber industry had been in decline since the recession of the mid-1980s. While the injunction to stop the felling of old-growth forest did lead to a short decline in lumber-related employment, the overall rate of decline would normalize later that year. See Josh Lehner, "Historical Look at Oregon's Wood Product Industry," Oregon Office of Economic Analysis, January 23, 2012, https://oregoneconomicanalysis.com/2012/01/23/historical-look-at-oregons-wood-product-industry/.

201 *Today these large swaths of old-growth forest* Ken Rait, "The Economic Value of 'Quiet Recreation' on BLM Lands," Pew Charitable Trusts, March 31, 2016, http://www.pewtrusts.org/en/research-and-analysis/analysis/2016/03/31/the-economic-value-of-quiet-recreation-on-blm-lands. Also see the report "West Is Best: How Public Lands in the West Create a Competitive Economic Advantage," Headwaters Economics, December 2012, https://headwaterseconomics.org/economic-development/trends-performance/west-is-best-value-of-public-lands/.

201 *$16 billion outdoor recreation industry* For more on Oregon's outdoor recreation industry, see "Oregon," Outdoor Industry Association, 2017, https://outdoorindustry.org/wp-content/uploads/2017/07/OIA__RecEcoState__OR.pdf.

203 *"conserving the stage"* To learn more about this conservation strategy, read Mark G. Anderson and Charles E. Ferree, "Conserving the Stage: Climate Change and the Geophysical Underpinnings of Species Diversity," *PLoS ONE* 5, no. 7 (July 14, 2010), https://doi.org/10.1371/journal.pone.0011554.

203 *species are on the move* Check out Jim Robbins, "Resilience: A New Conservation Strategy for a Warmer World," *Yale Environment 360* (July 13, 2015), http://e360.yale.edu/features/resilience__a__new__conservation__strategy__for__a__warming__world.

203 *at a respective rate of forty vertical feet and eleven miles every decade* Sara Reardon, "In Warming World, Critters Run to the Hills," *Science*, August 18, 2011, http://www.sciencemag.org/news/2011/08/warming-world-critters-run-hills.

203 *Douglas fir–dominated old-growth stands* For more information on the *Pseudotsuga menziesii* (or the coast Douglas fir), see Richard K. Hermann and Denis P. Lavender, "Douglas-Fir," US Forest Service, https://www.srs.fs.usda.gov/pubs/misc/ag__654/volume__1/pseudotsuga/menziesii.htm.

204 *roughly half of the over fourteen hundred plants and animals* See note on endangered species and wetlands dependence for page 6, as well as the South Bay Salt Pond Restoration Project's Fact Sheet, http://www.southbayrestoration.org/Fact%20Sheets/FS2.html. Also see S. L. Pimm et al., "The Biodiversity of Species and Their Rates of Extinction, Distribution, and Protection," *Science* 344, no. 6187 (May 2014), https://doi.org/10.1126/science.1246752.

On Restoration

Transcribed from an in-person interview with Richard Santos at Taco del Oro in Alviso, California, on April 18, 2017, and a telephone interview on September 12, 2017.

Looking Backward and Forward in Time

215 *While the indigenous people of California* Much of the historic information on the Bay Area's wetlands comes from Matthew Booker's precise, exhaustively researched, and

insightful *Down by the Bay: San Francisco's History between the Tides* (Berkeley: University of California Press, 2013). If only Matthew could have written this kind of investigation of each and every single area I have covered in *Rising*, my job would have been ten—if not a hundred—times easier. My debt to his work in this chapter is great.

215 *encompasses about fifteen thousand of the forty thousand acres of former tidelands* See the South Bay Salt Pond Restoration Project website: http://www.southbayresto-ration.org/Project__Description.html.

217 *And then I found Measure AA* For more on this groundbreaking ballot measure, check out John Upton, "Bay Area Voters Approve Tax to Fix Marshes As Seas Rise," Climate Central, June 8, 2016, http://www.climatecentral.org/news/bay-area-voters-approve-tax-fix-marshes-seas-rise-20420. Also please consult the Measure's online entry in Ballotpedia: https://ballotpedia.org/San__Francisco__Bay__Restoration__Authority__%E2%80%9CClean__and__Healthy__Bay%E2%80%9D__Parcel__Tax,__Measure__AA__(June__2016).

217 *This is how Hoover* This section draws on information from two sources (in addition to population data), one exhaustive, the other immersive: Marc Reisner, *Cadillac Desert: The American West and Its Disappearing Desert* (New York: Penguin, 1986) and David Owen, "Where the River Runs Dry," *New Yorker*, May 25, 2015, https://www.newyorker.com/magazine/2015/05/25/the-disappearing-river.

219 *and between the Columbia River and Baja* Matthew Booker, *Down by the Bay*.

219 *prior to the arrival of Spanish missionaries* Chuck Striplen, "Indigenous Tribes and Languages of the San Francisco Bay Area," pamphlet, *Fisher Bay Observatory Essays/ San Francisco Exploratorium*, March 21, 2016. You can find the map that accompanies this essay here: http://www.sfei.org/news/native-languages-map-bay-explor-atorium#sthash.PEe2AM78.dpbs.

219 *they harvested oysters and clams* Matthew Booker, *Down by the Bay*.

220 *nearly 790,000 acres of California's wetlands* Ibid.

220 *shapes so unlike their old selves* Thanks to Ben Pease, the cartographer behind "Once and Future Waters," a map in *Infinite City: A San Francisco Atlas*, ed. Rebecca Solnit (Berkeley: University of California Press, 2010). Pease made all of the maps in that book, including the one which accompanies the essay I mention later, "The Names before the Names."

221 *water-hungry crops* Check out the neat infographic *National Geographic* put together in 2015 titled "Which California Crops Are Worth Their Weight in Water?": https:// news.nationalgeographic.com/2015/05/150508-which-california-exports-crops-are-worth-the-water/.

222 *"The Army Corps of Engineers came up"* The full report is over one thousand pages long! It is titled "South San Francisco Shoreline Study Final Integrated

Interim Feasibility Study and Environmental Impact Statement/Environmental Impact Report." You can read the entire thing here: http://www.spn.usace.army. mil/Portals/68/docs/FOIA%20Hot%20Topic%20Docs/SSF%20Bay%20 Shoreline%20Study/Final%20Shoreline%20Main%20Report.pdf.

224 *This is how NeilArmstrong* There are numerous videos on YouTube that document the training the Apollo 11 astronauts went through to prepare for man's first trip to the moon. Also check out Mark Wolverton, "The G Machine," *Air & Space*, May 2007, https://www.airspacemag.com/history-of-flight/the-g-machine-16799374/, and "Fact Sheets: NASA Langley Research Center's Contributions to the Apollo Program," https://www.nasa.gov/centers/langley/news/factsheets/Apollo.html.

225 *for the overwhelming majority* This information comes from Matthew Booker, *Down by the Bay*, and Lisa Conrad's essay "The Names before the Names" in *Infinite City*.

226 *This is how Frank Oppenheimer* Much of the background information comes from Susan Schwartzenberg, and what little I have added comes from an interview conducted by Ira Flatow with K. C. Cole, the author of *Something Incredibly Wonderful Happens: Frank Oppenheimer and the World He Made Up* (New York: Houghton Mifflin Harcourt, 2009), a biography of the younger of the two famous Oppenheimer physicists: "Exploratorium Founder Profiled in New Book," National Public Radio, August 7, 2009, https://www.npr.org/templates/story/story.php?storyId=111658432.

226 *"inexcusable . . . reprehensible"* For a quick and dirty introduction to the rise and fall of Joseph McCarthy, see the History Channel's introduction to the man here: "Joseph McCarthy," 2009, https://www.history.com/topics/cold-war/joseph-mccarthy. Interested in a deeper dive? The Senate recently published the transcripts of the Army-McCarthy hearings (all well over three thousand pages of them) online here: https://www.senate.gov/artandhistory/history/common/generic/ McCarthy__Transcripts.htm.

229 *much of North America was covered* For more information on historic sea level rise along the Pacific coast of the United States, see the National Research Council's report "Sea-Level Rise for the Coasts of California, Oregon, and Washington: Past, Present, and Future," 2012, https://www.nap.edu/read/13389/chapter/3#14.

230 *In the time it has taken me to write this book* Compare, for instance, "Sea-Level Rise for the Coasts of California, Oregon, and Washington: Past, Present, and Future"—a 2012 report from the Committee on Sea Level Rise in California, Oregon, and Washington, National Research Council—with the follow-up report they released in 2017. Also see this new study of sea level rise rates, which relies on observational data to suggest that the total predicted rise by 2100 ought to be doubled: R. S. Nerem et al., "Climate-Change-Driven Accelerated Sea Level Rise Detected in the Altimeter Era."

231 *This is how Frank's brother Robert* Most of this information comes from Richard Rhodes, "The Atomic Bomb and Its Consequences" (lecture, Hanford Site, Washington), reprinted in James Conca, "Why Did We Make the Atomic Bomb?," *Forbes*, December 7, 2013, https://www.forbes.com/sites/jamesconca/2013/12/07/

why-did-we-make-the-atomic-bomb/#760442bd6e90. Other information comes from a primary document listing fatalities at Los Alamos, which can be found in *Los Alamos Project Y, Book II: Army Organization, Administration, and Operation*, copy in *Manhattan Project: Official History and Documents* [microform] (Washington, DC: University Publications of America, 1977), reel 12. Also see Alex Wellerstein, "The Demon Core and the Strange Death of Louis Slotin," *New Yorker*, May 21, 2016, https://www.newyorker.com/tech/elements/demon-core-the-strange-death-of-louis-slotin. Finally, check out the Atomic Heritage Foundation's website for an amazing directory of all of the different people employed during the quest to build a nuclear warhead: https://www.atomicheritage.org/.

233 *"some of the most valuable per square foot"* This sentiment was shared by Will Travis, the former director of the San Francisco Bay Area Conservation and Development Commission, in an interview on April 19, 2017, at a Starbucks in Berkeley, and by Dave Pine, the District One supervisor in San Mateo County, and his staff in a formal meeting on April 18, 2017, in the county offices. In both instances, both men were as surprised to hear about retreat as John was.

235 *This is how Harriet Tubman* Most of these facts are sourced from "Harriet Tubman," *Africans in America* Resource Bank, PBS, http://www.pbs.org/wgbh/aia/part4/4p1535.html, and the Harriet Tubman Historical Society, at http://www.harriet-tubman.org/. The rest come from Catherine Clinton, *Harriet Tubman: The Road to Freedom* (New York: Back Bay Books, 2005).

240 *This is how Robert Moses* The information in this section is gathered from many resources, chief among them Robert Caro's Pulitzer Prize–winning tome, *The Power Broker: Robert Moses and the Fall of New York* (New York: Vintage, 1975). Also worthwhile are Matthew Power, "The Cherry Tree Garden: A Rural Strongold in South Bronx," *Granta* (May 2008), and Robert Caro, "Robert Caro Wonders What New York Is Going to Become," interview by Christopher Robbins, *Gothamist*, February 17, 2016, http://gothamist.com/2016/02/17/robert__caro__author__interview.php. The information about New York City's segregated school system comes from John Kucsera, "New York State's Extreme School Segregation: Inequality, Inaction, and a Damaged Future," Civil Rights Project, UCLA, March 26, 2014, https://www.civilrightsproject.ucla.edu/research/k-12-education/integration-and-diversity/ny-norflet-report-placeholder.

242 *As I walk I am thinking* Thank you to Ta-Nehisi Coates's *Between the World and Me* (New York: Spiegel & Grau, 2015), the litany approach of which inspired this section, and the content of which kept equity in the front of my mind while I was writing.

242 *environmental impact report* "Environmental Impact Report—Facebook Campus Project," City of Menlo Park, https://www.menlopark.org/648/Environmental-Impact-Report.

244 *Facebook just constructed a 430,000-foot campus* Oliver Milman, "Facebook, Google Campuses at Risk of Being Flooded Due to Sea Level Rise," *The Guardian*, April 22, 2016, https://www.theguardian.com/technology/2016/apr/22/silicon-valley-sea-level-rise-google-facebook-flood-risk. Also see the Instagram hashtag

#mpk2ofirstlook, a collection of photographs staffers posted in the building's first days. The total cost of the building was estimated via Menlo Park building permit records in 2013, 2014, 2015, and 2016.

246 *I visit Oro Loma* Oro Loma *is* in fact promising and progressive. To learn more about it, check out their website: https://oroloma.org/horizontal-levee-project/.

Listening at the Water's Edge

254 *79 percent of participants* Devon McGhee, "Quantifying the Success of Buyout Programs: A Staten Island Case Study," CAKE Case Studies (2017), https://www.cakex.org/case-studies/quantifying-success-buyout-programs-staten-island-case-study.

255 *"We tell ourselves stories"* Joan Didion, "The White Album," in *The White Album: Essays* (New York: Farrar, Straus & Giroux, 1979).

256 *what Amitav Ghosh calls* Amitav Ghosh, *The Great Derangement: Climate Change and the Unthinkable* (Chicago: University of Chicago Press, 2016).

256 *This realization brought* This move toward collective action has begun to play out on the national level. On the eve of the single costliest hurricane season in United States history, Harriet Festing founded Flood Forum USA (www.floodforum. org), the first nationwide coalition designed to help flood survivors get organized, heard, and supported. Flood Forum—which has, in the year since its founding, already attracted more than twenty-five thousand members—plays matchmaker, connecting local activists to the support they need in order to address flooding in an equitable manner. They introduce frontline communities to scientists, designers, and lawyers who are willing to donate their time and expertise to help residents understand their flood risk, communicate those risks to local representatives, propose long-term solutions, fight future wetland development, and hold developers who have unlawfully built in wetlands financially accountable for their actions.

257 *"the magnitude and interconnectedness"* Ghosh, *The Great Derangement*.

259 *"I experienced not"* Candis Callison, *How Climate Change Comes to Matter: The Communal Life of Facts* (Durham, NC: Duke University Press, 2014).

260 *from the tens of millions of indigenous peoples killed* Estimates of the precontact population of the Americas range significantly from as little as 8 million people to up to 112 million, the latter being one of the more recent estimates. All calculations are estimates and so I have chosen somewhat vague language that accommodates the bulk of the proposals that have been put forth—and yet I would also not be surprised if our estimates continue to rise in the coming years. See: *The Native Population in the Americas in 1492*, ed. William M. Denevan (Madison: University of Wisconsin Press, 1992).

261 *"landscaped berms with sea grass"* "Rebuild by Design, The Big U," 2016 ASLA Professional Awards, American Society of Landscape Architects, https://www. asla.org/2016awards/172453.html.

261 *only 37 percent* Greg Allen, "Ghosts of Katrina Still Haunt New Orleans' Shattered Lower Ninth Ward," National Public Radio, August 3, 2013, https://www. npr.org/2015/08/03/427844717/ghosts-of-katrina-still-haunt-new-orleans-shattered-lower-ninth-ward.

262 *For those who remained, resilience might mean* Gary Rivlin, "Why the Lower Ninth Ward Looks Like the Hurricane Just Hit," *The Nation*, August 13, 2015, https://www.thenation.com/article/why-the-lower-ninth-ward-looks-like-the-hurricane-just-hit/.

262 *New Orleans's "revival"* Judith Rodin, "The Secret to New Orleans' Comeback," *Fortune*, September 3, 2015, http://fortune.com/2015/09/03/secret-new-orleans-comeback/.

262 *a map that marked* For more on the history and reception of the "green dot" map, see Richard Campanella's "A Katrina Lexicon," *Places Journal*, July 2, 2015, https:// placesjournal.org/article/a-katrina-lexicon/.

263 *they had been forced to occupy* Josh Lewis and Henrik Ernstson, "Contesting the Coast: Ecosystems as Infrastructure in the Mississippi River Delta," *Progress in Planning*, December, 1, 2017, https://doi.org/10.1016/j.progress.2017.10.003.

Stephanie Alvarez Ewens

ELIZABETH RUSH's work has appeared in the *New York Times*, the *Washington Post*, *The Guardian*, *The Atlantic*, *Harper's*, *Pacific Standard*, and the *New Republic*, among many others. She is the recipient of fellowships and grants including the Howard Foundation Fellowship, awarded by Brown University; the Andrew Mellon Foundation Fellowship for Pedagogical Innovation in the Humanities; the Metcalf Institute Fellowship; and the Science in Society Journalism Award from the National Association of Science Writers. She received her MFA in nonfiction from Southern New Hampshire University and her BA from Reed College. She lives in Rhode Island, where she teaches creative nonfiction at Brown University.

milkweed
editions

Founded as a nonprofit organization in 1980, Milkweed Editions is an independent publisher. Our mission is to identify, nurture and publish transformative literature, and build an engaged community around it.

Milkweed Editions is based in Bdé Óta Othúŋwe (Minneapolis) within Mní Sota Makhóčhe, the traditional homeland of the Dakhóta people. Residing here since time immemorial, Dakhóta people still call Mní Sota Makhóčhe home, with four federally recognized Dakhóta nations and many more Dakhóta people residing in what is now the state of Minnesota. Due to continued legacies of colonization, genocide, and forced removal, generations of Dakhóta people remain disenfranchised from their traditional homeland. Presently, Mní Sota Makhóčhe has become a refuge and home for many Indigenous nations and peoples, including seven federally recognized Ojibwe nations. We humbly encourage our readers to reflect upon the historical legacies held in the lands they occupy.

milkweed.org

We are aided in this mission by generous individuals who make a gift to underwrite books on our list. Special underwriting for *Rising* was provided by the Hlavka family.

Interior design & composition by Mary Austin Speaker
Typeset in Vendetta

Vendetta was designed in 1999 by John Downer for the Emigre type foundry. The design of Vendetta was influenced by the design of types by roman punchcutters who traced their aesthetic lineage to Nicolas Jenson's seminal 1470 text, *De Evangelica Praeparatione*, a work of Christian apologetics written in the fourth century AD by the historian Eusebius.